Studies in Systems, Decision and Control

Volume 412

The series "Studies in Systems, Decision and Control" (SSDC) covers both new developments and advances, as well as the state of the art, in the various areas of broadly perceived systems, decision making and control–quickly, up to date and with a high quality. The intent is to cover the theory, applications, and perspectives on the state of the art and future developments relevant to systems, decision making, control, complex processes and related areas, as embedded in the fields of engineering, computer science, physics, economics, social and life sciences, as well as the paradigms and methodologies behind them. The series contains monographs, textbooks, lecture notes and edited volumes in systems, decision making and control spanning the areas of Cyber-Physical Systems, Autonomous Systems, Sensor Networks, Control Systems, Energy Systems, Automotive Systems, Biological Systems, Vehicular Networking and Connected Vehicles, Aerospace Systems, Automation, Manufacturing, Smart Grids, Nonlinear Systems, Power Systems, Robotics, Social Systems, Economic Systems and other. Of particular value to both the contributors and the readership are the short publication timeframe and the world-wide distribution and exposure which enable both a wide and rapid dissemination of research output.

Indexed by SCOPUS, DBLP, WTI Frankfurt eG, zbMATH, SCImago.

All books published in the series are submitted for consideration in Web of Science.

More information about this series at https://link.springer.com/bookseries/13304

Aboul Ella Hassanien · Deepak Gupta ·
Ashish Khanna · Adam Slowik
Editors

Virtual and Augmented Reality for Automobile Industry: Innovation Vision and Applications

Springer

Editors
Aboul Ella Hassanien
Faculty of Computer and AI
Cairo University
Giza, Egypt

Deepak Gupta
Department of Computer Science
and Engineering
Maharaja Agrasen Institute of Technology
Rohini, Delhi, India

Ashish Khanna
Department of Computer Science
and Engineering
Maharaja Agrasen Institute of Technology
Rohini, Delhi, India

Adam Slowik
Department of Electronics and Computer
Science
Koszalin University of Technology
Koszalin, Poland

ISSN 2198-4182 ISSN 2198-4190 (electronic)
Studies in Systems, Decision and Control
ISBN 978-3-030-94101-7 ISBN 978-3-030-94102-4 (eBook)
https://doi.org/10.1007/978-3-030-94102-4

This Springer imprint is published by the registered company Springer Nature Switzerland AG
The registered company address is: Gewerbestrasse 11, 6330 Cham, Switzerland

Preface

The Augmented Reality (AR) and Virtual Reality (VR) automotive applications will increase road safety, bring intuitive activities to driving, and finally enhance driving experience. AR/VR technology may also help on the transition toward automated driving. AR Head-Up-Displays (HUDs) may soon overlay 3D navigation instructions onto road geometry and moving obstacles like vulnerable road users (pedestrians, bikers, wheel-chair users) and other vehicles may be highlighted to calm down the driver-passenger and enhance trust in their vehicle's automated operation as the vehicle proves its awareness of its surroundings. VR windshields may allow for dynamic reconfiguration of multi-lane roads based on demand and will, in the long term, remove road signs, traffic lights, road paintings, etc. from the streets.

However, many technological challenges need to be addressed before AR/VR applications will hit the mainstream market. These include how to capture and interpret road geometry through computing intensive sensor fusion, precise vehicle positioning, compensation for vibrations, delays, and jitter, laser projection, driver monitoring via inward facing cameras and designing sophisticated algorithms to generate precise augmentation content in the viewing field of the driver, etc.

This book presents the Augmented Reality (AR) and Virtual Reality (VR) automotive applications. It unites automobile with a leading technology, i.e., augmented and virtual reality and uses the advantages of the latter to solve the problems faced by the former. The book highlights the reasons for the growing abundance and complexity in this sector. Virtual and augmented reality presents a powerful engineering tool that finds application in various engineering fields. The book further illustrates the possible challenges in its applications and suggests ways to overcome them. The book includes nine chapters focusing on automobile collision avoidance, self-driving cars, autonomous vehicles, navigation systems, and many more applications.

The topic is wide in nature, and hence every technique and/or solution cannot be discussed in detail. The primary emphasis of this book is to introduce autonomous industry, innovation, vision, and applications to data scientists, students, and academicians at large.

Objective of the Book

The main aim of this book is to provide a detailed understanding of Virtual and Augmented Reality and focus on Automobile Industry: Innovation Vision and Applications.

Organization of the Book

The book is organized in nine chapters with the following brief description:

Automotive Collision Avoidance System: A Review

A Vehicle Collision Avoidance System aims to develop safety features in automobiles that prevent collisions with other vehicles or obstacles in the road. These technologies are convenient in inclement weather. When driving, the main goal of these devices is to prevent car collisions caused by negligence or blind spots. As a result, the focus of this review paper is on vehicle accident prevention systems.

Motion planning and Maneuvering in Self-Driving Cars

The purpose of this study is to review present approaches and evaluate the different techniques for the motion planning of self-directed on-road driving that consists of locating a lane, finding for the safest maneuver, and shaping the main possible route. The study demonstrates a critical evaluation of each of these processes in terms of their benefits and drawbacks.

Autonomous Vehicle Assisted by Heads up Display (HUD) with Augmented Reality Based on Machine Learning Techniques

Machine learning and deep neural networks are used in this chapter for lab and real-world T&V for ARHUD and autonomous vehicles. The results of simulation obtain the data gathered from the implementation of Human and Machine Interface (HMI) to detect the object and to classify the objects in motion. The simulation results obtained are accuracy of 98%, precision 94%, recall 92.3%, and F-1 score 86% in comparison with CNN, ANN, and SVM.

Special Sensors for Autonomous Navigation Systems in Crops Investigation System

A summary of sensor technology for autonomous prototype vehicles is presented and discussed in this chapter. Navigating sensors, computer methods, and navigation management approaches are the main aspects. Crucial procedures include selecting, coordinating, and combining the appropriate sensors to provide essential robotics navigational knowledge. For function extraction, processing of data and fusing computationally efficiently are utilized. In this chapter, we will discuss about special sensor keep monitoring through GPS system requirement of crops and to improve and fine growth of quality seeds.

AR/VR Technology for Autonomous Vehicles and Knowledge-Based Risk Assessment

The purpose of this chapter is to estimate the danger of adopting AR/VR technology in the automotive industry since real-time risk assessment of autonomous driving at operational and tactical levels is exceedingly difficult because both contextual and circumferential aspects must be considered simultaneously.

Optimal Stacked Sparse Autoencoder-Based Traffic Flow Prediction in Intelligent Transportation Systems

This study designs an optimal stacked sparse autoencoder-based traffic flow prediction (OSSAE-TFP) model for ITS. The goal of the OSSAE-TFP technique is to determine the level of traffic flow in ITS. In addition, the presented OSSAE-TFP technique involves the traffic and weather data for TFP. To showcase the enhanced predictive outcome of the OSSAE-TFP technique, a wide range of simulations was carried out on benchmark datasets and the results portrayed the supremacy of the OSSAE-TFP technique over the recent state-of-the-art methods.

Leverage Computer Vision for Cost-Effective Learning Paradigm

In this chapter, a model has been proposed which facilitated the user to explore different dimensions of the computer vision which are required for understating various concepts in a very cost-effective manner. In this model, a vision-based interface for the user replaces peripheral devices without affecting their performance and makes it easier for them to use these interfaces built into AR to better understand 3D concepts.

Hand Gesture Recognition for Real-Time Game Play using Background Elimination and Deep Convolution Neural Network

The novelty of work is to play games using any good quality mobile/web camera instead of the game controller. A Convolution Neural Network (CNN) is combined with background elimination to detect different hand gestures and further uses these gestures to control video games. The designed human-computer interaction system concludes 98.2% accuracy of hand gesture recognition for considered hand gestures. The measured latency is also adequate while game play.

Modeling of Optimal Bidirectional LSTM-Based Human Motion Recognition for Virtual Reality Environment

This study designs a new optimal Bidirectional Long Short-Term Memory (BiLSTM) with Fully Convolution Network (FCN) called OBiLSTM-FCN model for human motion recognition in VR environment. The proposed OBiLSTM-FCN model comprises different processes, namely, feature extraction, classification, and hyper-parameter optimization. The performance validation of the OBiLSTM-FCN model takes place and the resultant values portrayed the improved performance over the other compared methods.

Giza, Egypt — Aboul Ella Hassanien
Rohini, India — Deepak Gupta
Rohini, India — Ashish Khanna
Koszalin, Poland — Adam Slowik

About This Book

This book presents the Augmented Reality (AR) and Virtual Reality (VR) automotive applications. It unites automobile with a leading technology, i.e., augmented and virtual reality and uses the advantages of the latter to solve the problems faced by the former. The book highlights the reasons for the growing abundance and complexity in this sector. Virtual and augmented reality presents a powerful engineering tool that finds application in various engineering fields. The book further illustrates the possible challenges in its applications and suggests ways to overcome them. The book includes nine chapters focusing on automobile collision avoidance, self-driving cars, autonomous vehicles, navigation systems, and many more applications.

Contents

About the Editors

Prof. Aboul Ella Hassanien is the Founder and Head of the Egyptian Scientific Research Group (SRGE) and a Professor of Information Technology at the Faculty of Computer and Artificial Intelligence, Cairo University. He is an ex-dean of the Faculty of Computers and Information, Beni Suef University. He has more than 800 scientific research papers published in prestigious international journals and over 40 books covering such diverse topics as data mining, medical images, intelligent systems, social networks, and smart environment. He won several awards, including the Best Researcher of the Youth Award of Astronomy and Geophysics of the National Research Institute, Academy of Scientific Research (Egypt, 1990). He was also granted a scientific excellence award in humanities from the University of Kuwait for the 2004 Award and received the scientific—University Award (Cairo University, 2013). Also, he was honored in Egypt as the best researcher at Cairo University in 2013. He also received the Islamic Educational, Scientific and Cultural Organization (ISESCO) prize on Technology (2014) and received the State Award for excellence in engineering sciences 2015. He was awarded the medal of Sciences and Arts of the first class by the President of the Arab Republic of Egypt, 2017.

Deepak Gupta [M'19, SM'20] received a B.Tech. degree in 2006 from the Guru Gobind Singh Indraprastha University, Delhi, India. He received M.E. degree in 2010 from Delhi Technological University, India and Ph. D. degree in 2017 from Dr. APJ Abdul Kalam Technical University (AKTU), Lucknow, India. He has completed his Post-Doc from National Institute of Telecommunications (Inatel), Brazil in 2018. He has co-authored more than 166 journal articles including 128 SCI papers and 51 conference articles. He has authored/edited 59 books, published by IEEE-Wiley, Elsevier, Springer, Wiley, CRC Press, DeGruyter and Katsons. He has filed four Indian patents. He is convener of ICICC, ICDAM, and DoSCI Springer conference series. Currently he is Associate Editor of Expert Systems (Wiley), and Intelligent Decision Technologies (IOS Press). He is the recipient of 2021 IEEE System Council Best Paper Award. He have been featured in the list of top 2% scientist/researcher database in the world [Table-S7-singleyr-2019]. He is also working toward promoting

Startups and also serving as a Startup Consultant. He is also a series editor of "Elsevier Biomedical Engineering" at Academic Press, Elsevier, "Intelligent Biomedical Data Analysis" at De Gruyter, Germany, "Explainable AI (XAI) for Engineering Applications" at CRC Press. He is appointed as Consulting Editor at Elsevier.

Ashish Khanna has 16 years of expertise in Teaching, Entrepreneurship, and Research and Development. He received his Ph.D. degree from National Institute of Technology, Kurukshetra. He has completed his M. Tech. and B. Tech. from GGSIPU, Delhi. He has completed his Post-Doc from Internet of Things Lab at Inatel, Brazil and University of Valladolid, Spain. He has published around 50 SCI indexed papers in IEEE Transaction, Springer, Elsevier, Wiley, and many more reputed journals with cumulative impact factor of above 100. He has around 100 research articles in top SCI/Scopus journals, conferences, and book chapters. He is co-author of around 20 edited and text books. His research interest includes distributed systems, MANET, FANET, VANET, IoT, machine learning, and many more. He is originator of Bhavya Publications and Universal Innovator Lab. Universal Innovator is actively involved in research, innovation, conferences, startup funding events, and workshops. He has served the research field as a Keynote Speaker/Faculty Resource Person/Session Chair/Reviewer/TPC member/post-doctorate supervision. He is Convener and Organizer of ICICC conference series. He is currently working at the Department of Computer Science and Engineering, Maharaja Agrasen Institute of Technology, under GGSIPU, Delhi, India. He is also serving as Series Editor in Elsevier and De Gruyter publishing houses.

Dr. Adam Slowik (IEEE Member 2007; IEEE Senior Member 2012) received the B.Sc. and M.Sc. degrees in computer engineering and electronics in 2001 and the Ph.D. degree with distinction in 2007 from the Department of Electronics and Computer Science, Koszalin University of Technology, Koszalin, Poland. He received the Dr. habil. degree in computer science (intelligent systems) in 2013 from the Department of Mechanical Engineering and Computer Science, Czestochowa University of Technology, Czestochowa, Poland. Since October 2013, he has been an Associate Professor in the Department of Electronics and Computer Science, Koszalin University of Technology. His research interests include soft computing, computational intelligence, and, particularly, bio-inspired optimization algorithms and their engineering applications. He is a reviewer for many international scientific journals. He is an author or co-author of over 80 refereed articles in international journals, two books, and conference proceedings, including one invited talk. He is an Associate Editor of the IEEE Transactions on Industrial Informatics. He is a member of the program committees of several important international conferences in the area of artificial intelligence and evolutionary computation. He was a recipient of one Best Paper Award (IEEE Conference on Human System Interaction—HSI 2008). He was a Guest Editor in Special Issues organized by such journals as IEEE Transactions on Industrial Electronics, IEEE Transactions on Industrial Informatics, and IEEE Transactions on Fuzzy Systems.

Automotive Collision Avoidance System: A Review

A. Rammohan, Suresh Chavhan, Ramesh Kumar Chidambaram, N. Manisaran, and K. V. Pavan Kumar

Abstract Vision-based vehicle identification strategies for improving road safety have gotten much attention in the last decade. Unfortunately, the techniques are not very robust due to detecting the wide range of vehicle shapes, crowded environments, different lighting situations, and driving behavior. Accidents are becoming increasingly common and unpredictable. Accidents result in the most severe damage, severe injury, and even death. The majority of these collisions are caused by the driver's failure to hit the brakes promptly. Preventive measures such as improved visibility, auto headlamps, windshield wipers, and tyre traction lessen the chances of getting into an accident. A Vehicle Collision Avoidance System aims to develop safety features in automobiles that prevent collisions with other vehicles or obstacles in the road. These technologies are convenient in inclement weather. While driving, the main goal of these devices is to prevent car collisions caused by negligence or blind spots. As a result, the focus of this review paper is on vehicle accident prevention systems.

Keywords Collision avoidance · Sensors · Safety · Collision warning · Sensor fusion

1 Introduction

Collision avoidance systems focus on cutting-edge concepts like pre-crash sensing. An ultrasonic sensor detects an object in front of the car and sends a signal to the microcontroller unit. The microcontroller unit delivers a signal to the braking unit based on the signal received from the ultrasonic sensor, causing the brake to be applied automatically. A collision avoidance system also called as pre-collision

A. Rammohan (✉) · S. Chavhan · R. K. Chidambaram
Automotive Research Centre, Vellore Institute of Technology, Vellore 632014, India
e-mail: rammohan.a@vit.ac.in

N. Manisaran · K. V. P. Kumar
Department of Automotive Engineering, Vellore Institute of Technology, Vellore 632014, India

A. E. Hassanien et al. (eds.), *Virtual and Augmented Reality for Automobile Industry: Innovation Vision and Applications*, Studies in Systems, Decision and Control 412, https://doi.org/10.1007/978-3-030-94102-4_1

system, forward-collision warning system, or active safety system, is a type of automotive safety technology designed to avoid or lessen the degree of a collision's severity. It detects an impending crash using radar (all-weather), LIDAR (light detection and ranging) with the camera. Through a location database, GPS sensors can identify stationary risks such as approaching stop signs.

When these systems detect an imminent accident, they issue a warning to the driver. When a collision is near, they respond independently, without the need for driver intervention (by braking or steering or both). At low vehicle speeds, avoidance of collisions braking is acceptable, but at a more incredible vehicle speed, collision avoidance by turning is possibly more appropriate if lanes. There are four different forms of fatigue.

- Localized physical tiredness
- General physical fatigue
- Central nervous fatigue
- Mental fatigue.

Insufficient knowledge is also a significant cause of traffic accidents, such as dangerous shortcut maneuvers that endanger opposing vehicles and late-night or early-morning inability to recognize traffic signals. This is especially dangerous for road riders at risks, such as Bikers, riders, and pedestrians. Motorcyclists have the highest percentage of accidents due to their lack of protection and high speeds. The ASEAN region has a higher accident rate than the rest of the world. A sudden impediment or a wheel slide can easily lead a biker to lose control in traffic, leading to an accident. Other factors for lane splitting include speeding, traveling on the opposite side of the road, driving while under the influence of narcotics, not using turn signals, vehicle (or motorbike) malfunctions, and deliberate, aggressive behavior are the examples of reckless driving.

The motivation of this review work is to encapsulate the techniques involved in the collision warning and avoidance systems. The main contribution of this study involves detailed study of various sensors used for collision avoidance such as active and passive, sensor fusion applicable for collision avoidance. The study also involved various detection techniques which is found ad important data set for calculating the collision time. The paper is organised as Literature survey for collision warning and avoidance, its associated sensors, various vehicle detection systems and challenges in vehicle detection.

2 Literature Survey

Collision warning and collision avoidance systems are new car safety technology that helps drivers prevent rear-end incidents. Their purpose is to provide the driver ample time to avoid a collision while avoiding bothering the driver with signals deemed too early or unnecessary. Connected vehicle (CV) technologies, in conjunction with other Driving Assistance (DA) technologies, are predicted to have a heavy impact on traffic operations and safety and the future of our cities. However, few studies have calculated the actual safety benefits of equipping all vehicles with this technology [1, 2]. The introduction of car collision warning systems could be the next significant step forward in automobile safety. These devices attempt to actively warn drivers of impending collisions, giving them enough time to take appropriate remedial action to lessen or prevent the incident. Active safety systems of this type will be effective, according to the fatality rate and analysis. The Collision Avoidance System has advanced to a new level of development, actively and considerably improving the safety of automobile passengers. However, research only looked at the time when vehicles are traveling, ignoring the time when passengers are most vulnerable—when they step out of the car and their attention to the surroundings lowers, and the vehicle's structure no longer shields them [3, 4].

Intelligent transportation system (ITS) is known as the application of smart technology to improve transportation systems by making autos and road traffic infrastructure intelligent to deliver a safe and comfortable environment. Sensors, message communication, and traffic management technology are all part of ITS [5]. Collison ultrasonic sound sensor technology was avoided. It recognizes incoming cars and displays information to the driver. The system will determine the distance between two vehicles traveling in the same lane and direction. The system activates safety measures on the vehicle when an object's trajectory is directed toward its vehicle and becomes dangerous [6].

A microprocessor and internal sensors are used to create a vehicle safety system. It primarily focuses on the use of the CAN protocol to construct a driver-vehicle interaction in digital format. By and large, the CAN architecture reduces wiring. Information such as humidity, temperature, and distance are acquired using this technique. The alert delivers a notification to the driver if an impediment is detected within 300 cm of the vehicle. When a car reaches a pollution zone, it slows down built on vehicle detection processes, which begin with sensor selection and end with vehicle identification and tracking [7]. Each process/technique step is examined and studied separately. The study's two main contributions are a survey on motorcycle detection methods and a sensor comparison regarding cost and range characteristics. Finally, the poll recommends a low-cost, dependable collision avoidance system for the automotive industry [8].

Distracted driving and inattention are the leading causes of car accidents, with the majority of them resulting in fatalities. It is essential to create information systems that can identify driver carelessness and diversion to minimize road accidents. Diversion monitoring systems for automobiles are currently either unavailable or confined to specific causes of driver carelessness, such as fatigue. Even though driving is becoming increasingly automated due to the Increasingly advanced help technologies available, automated cars will continue to play a more significant role as car technology advances [9].

Immature driving behaviour, such as shortcut manoeuvres that endanger opposing vehicles or late at night or early in the morning, inability to follow traffic signals, is a significant contributor to road accidents. This is especially dangerous for rash road riders such as bikers, motorcyclists, and pedestrians. Motorcyclists have the highest number of accidents due to their lack of protection, and rapid speeds, which leads to traffic accidents that can be caused by various factors, including human mistakes and road and environmental conditions. Inadequate street lighting and climatic circumstances, such as foggy and wet weather, diminish visibility and make roadways slick. These are examples of the latter. Sharp turns, intersections, and junctions can all be found under the former category. Intersections account for almost one-third of all accidents [10, 11].

Vehicles must be equipped with safety equipment such as active or passive safety mechanisms are available. Seatbelts, head restraints, and side-impact are all standard equipment; for example, they were widely circulated used for many years and have nearly realized their total capacity in lowering the number of fatalities [12]. Using strong duality of convex optimization, this study proposes a new technique for precisely active safety constraints that are non differentiable are restated as smooth, variational constraints [13].

Collision avoidance safety systems gained more significant study curiosity than autonomous collision avoidance systems since they assist in avert collisions in the event of close conditions. Nevertheless, because of issues with reliable environmental detection, Maintaining the resilience and dependability of collision avoidance aid and accurate collision warning strategic thinking and driver acceptance is tough. We looked at safety systems from the perspective of "collision avoidance". Headway distance measuring utilizing laser radar scan, Vehicle dynamics are included in the route estimate process, collision estimation of the automobile in front using a secure or dangerous judgment, and automatic braking system are among the system's core technological aspects [14, 15].

Intelligent vehicle technologies, mainly automated safety and accident-avoidance solutions, have received much attention in recent decades. A situation evaluation is critical for a successful car safety system in complex traffic settings. Figure 1 shows a diagram of the human-driving and ADAS processes. The human-driving process includes submodules such as the vision, purpose, and behavior of the driver. The advanced driver assistance system include functions such as Prediction & Detection and Threat estimation Decision-making. One of the most significant advancements in vehicle automation is the development of a completely autonomous car. Many studies have been conducted on vehicle automation, and one of the essential components of a completely driverless car is Advanced Driver Assistance Systems (ADAS). Automatic Cruise Control, Automatic Parking, Collision Avoidance, and Lane Departure Warning are standard ADAS features [16, 17].

Collision warning and collision avoidance systems are new car safety features that help drivers prevent rear-end incidents. They aim to give the driver enough time to avoid a collision without annoying them with too soon or unneeded alerts. The system that avoids collision gained more significant research curiosity than autonomous collision avoidance systems as they assist in avoiding near-collisions. Nevertheless, maintaining the robustness and reliability of collision avoidance assistance is difficult because of issues with reliable environmental detection, accurate collision avoidance decision making, and driver acceptability [18, 19].

Fig. 1 Vehicle collision avoidance system

Author presents a brief overview of the technologies proposed for predicting vehicle collisions before they occur and Intelligent safety mechanism operation such as the release of airbags and seat belt stiffening. The Internet of Things (IoT) collects and exchanges data between gadgets, cars, and other embedded objects. The internet of things has made the intelligent transport system (ITS) the solution for vehicle navigation, traffic management, Number plate recognition systems. It has progressed to implementations that provide real-time data and feedback from various sources, including weather conditions, vehicle parking suggestions, and traffic conditions. To prevent accidents at intersections, The proposed system focuses on a variable traffic signal sequence and depends on traffic density [20, 21].

In the field of vehicle research, collision avoidance has gotten much attention. Collision avoidance systems are classified as either sensor or ITS. The primary purpose of the research is to give a complete assessment of collision avoidance systems that use both sensors and ITS [22].

Driving is always fraught with danger. Various methods for reducing the risk have been presented. One of the most effective techniques for reducing accidents is critical motion detection of surrounding moving vehicles. The dynamic visual model (DVM) was introduced in this study as a computational model for detecting crucial motions of neighboring automobiles while driving on a highway. The DVM is inspired by the human visual system and comprises three types of analyzers: sensory analyzers, perceptual analyzers, and conceptual analyzers [23].

Avoiding collisions is a primary concern for most automobile systems. In an automobile system, a "Vehicle Collision Avoidance System" is a system of defense aimed towards lowering the possibility of a collision. The "Vehicle Collision Avoidance System" is utilized to track objects and take whatever necessary action to avoid a collision. These systems warn the driver when a collision has been detected, either by braking or steering. Moreover, the frequency of accidents in today's world is relatively high and unpredictable. Accidents result in the most severe damage, severe injury, and even death. The majority of these collisions are caused by the driver's failure to hit the brakes promptly. Preventive measures such as improved visibility, auto headlamps, windshield wipers, and tire traction reduce the chances of an accident. We have progressed to the point where we actively avoid collisions and give car occupants and even pedestrians [24, 25].

Tracking takes advantage of data continuity over time and enhances vehicle detection accuracy and speed significantly. It has two primary functions: estimating vehicle motion and position and maintaining track of automobiles that were previously recognized but were not detected in a specific frame [26, 27].

Human behavior is the primary cause of accidents. Vehicles need to be equipped with safety features. Active or passive safety mechanisms are available. Fatigue is the second most crucial element, accounting for around 25–30% of all road accidents. Mental and central nerve weariness are the most dangerous types of fatigue to drive with since they lead to drowsiness, which increases the risk of an accident [11, 28, 29].

Automobile collision avoidance (CA) systems assist drivers in avoiding accidents by stopping or steering autonomously. If the choice to interfere is taken too soon, the driver may find it inconvenient, and if the choice is made too late, the intervention's protection benefits will be diminished. A hazard function is frequently used to determine whether or not to act. The complexity of the danger function's input state space is generally very high, making comprehensive evaluation in actual vehicles impossible [30].

The subject of future customer adoption is critical for the proper usage of driver-assist systems, as it is for many breakthrough gadgets that need a unique user-technology interaction (ADAS). However, because adoption can only be evaluated after the technology has been used, data encouraging its usage should focus on enhancing early acceptability before using it. Acceptability refers to how people feel about a new product before they use it, and it can be impacted by the information they learn about it [31].

Automatic Emergency Braking (AEB) systems are increasingly being used in cars. These technologies are designed to assist the driver in avoiding or minimizing crashes by hitting the brakes immediately before a collision occurs. Earlier, only rear-end collisions were handled, but new collision types have been included as time has gone on. Brakes have been deployed sooner and stronger to reduce velocity just before an accident happens [32].

Chain/cooperative Collision Avoidance (CCA) is a basic form of automotive wireless networks safety application (VANETs). They equip cars with collaborating communications networks to minimize the frequency of accidents. The most common method for evaluating these technologies is simulation. Present simulation techniques based on car-following models, on the other hand, cannot be utilized to mimic accidents perfectly [33].

Upcoming safety systems are the subject of a significant amount of study. Although much research is being done on innovations like collision avoidance systems, their use in solving high-risk crash situations every day on European roads and its practicality is still to be proven. The European Commission's eSafety initiative studies collision warning and collision avoidance systems in Japan, the United States, and the European Union. During scientific experiments, very significant estimations of such systems' safety capability have been stated. However, the variety of technological and behavioral difficulties associated with many of the ideas need a complete on-road evaluation. Most suggested solutions require an excellent traffic environment, such as highways, but with a limited potential for fatality reduction [34].

3 Automotive Collison Avoidance System

In today's world, traffic accidents are one of the leading causes of death and injury. Automobile manufacturers have begun to incorporate a growing number of driver assistance technologies to aid in preventing collisions. Adaptive cruise control (ACC), which is already available for numerous automobile types, is the initial stage towards CA systems for automotive applications. ACC systems adjust the speed if any in-path vehicle travels slower than the host vehicle's predetermined speed. The cruise control system can only apply a certain amount of deceleration (usually 3 m/s^2), and some systems will notify the driver if this acceleration is insufficient to avoid a collision. Current ACC systems are marketed as a convenience feature that the driver can turn on and off; they also disengage at low speeds (under 40 km/h). The next stage in automotive CA is always to provide active systems, execute autonomous braking, and warn of impending collisions. Even in typical driving, the traffic environment can be pretty complicated from a sensing standpoint. With various impediments to be identified and classified, it is a significant challenge in car accident avoidance. Furthermore, sensors and processing resources must be inexpensive.

4 Sensors for Collision Avoidance System

Detection of vehicles using vision, a quick discussion of other modalities. It is required to use technologies that are already used during on-road vehicle detection. As shown in Fig. 2, sensors in collision avoidance safety systems gather data on road conditions and are divided into two categories which are active and passive.

To identify targets/obstacles, the sensor sends out signals and detects the reflection signal. Later visual sensors or cameras, collect data in a non-intrusive manner.

Using multiple sensing technologies is a few method for assisting drivers in avoiding incidents and reducing traffic fatalities. As with rear backup cameras, basic systems send information to the driver. Others, park assist and lane change alerting

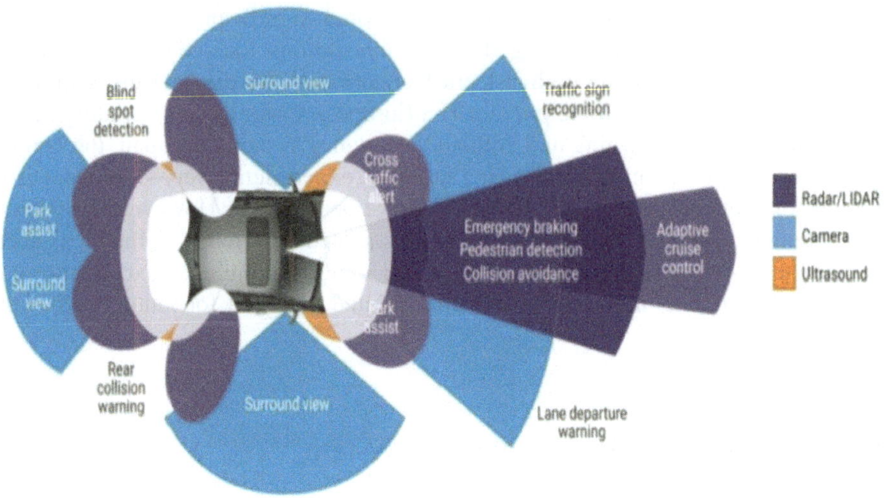

Fig. 2 Sensors in Vehicle collision avoidance system

systems, create an alert when an accident is imminent. To avoid immediate hazards, knowledgeable systems can alter the host vehicle's speed.

When a subject of attention is spotted, the system decides that the car is in a safe or risky region and alerts a potential collision. In situations, the system can also take control of the vehicle. Much research is currently being done in merging data from many sensors aboard the vehicle to feed a central fusion algorithm in charge of the complete collision avoidance system.

4.1 Active Sensors

Radar and Pulse-Doppler based on laser and Lidar (Light Detection and Ranging) are the most used active sensor techniques for detecting automobiles. To identify and track impediments in front of the vehicle, a radar framework was used. It was usually installed in the bottom front section of any vehicle. By monitoring reflections of radar signals, the system determines the separation between each vehicle and the objective and the respective speed. Using discrete-time signal processing, in real-time multiple-lane vehicle detection, radar-based driving safety features were demonstrated. The electromagnetic spectrum's such as ultraviolet, visible, and infrared waves are transmitted and received by laser and lidar-based systems.

A telescope is used to gather the waves that return to the receiver and count them as a function of time. Using the light speed, we can estimate how much the photons have traveled round trip. Lidar sensors, both 1D and 2D, are less expensive to manufacture. A laser is a more upgraded version to check laser range that determines the distance to an object using the time-of-flight principle. A laser scanner installed on an automobile can be used to detect and classify several vehicles. The calculated confidence level took the taking into account geometrical parameters, classifications, and time tracking. For improved accuracy, the device was sorely tested. in various settings) using three different lasers. A recent two-dimensional laser scanner mounted on the front end of a car was demonstrated in a test. Velodyne has recently released the HDL-64E, a compact 3D lidar sensor for obstacle detection and navigation in autonomous ground vehicles. It measures the environment with 64 fixed-mounted lasers, each mechanically attached at a precise angle, and the entire unit spinning. This method dramatically improves point cloud density, dependability, and field of view.

Active sensors are highly beneficial for identifying virtual environments and demonstrating fortitude in wet and foggy environments. The key benefit is that specific specified parameters (such as relative speed and distance) may be measured without complex mathematics. In snow, cloudy, and wet conditions, radar-based systems' long-range detection (about 150 m) is more reliable than lidar.

On the other hand, a typical lidar is less expensive than radar but has a lower range. Current lidar systems operate at high speeds that obtain high-resolution, and three-dimensional data are more expensive than radar, but they provide superior accuracy. These high-speed scanners can capture the outline of the target vehicle and classify it as a car, bus, truck, or motorcycle, among other things. Radar and acoustic sensors, on the other hand, Due to the changing and uncertain traffic congestion environments, less information about the target (such as shape, size, and colour) is more prone to interference concerns. Though prototype vehicles with active sensors produce better results when many cars travel the route, interference between similar sensors causes problems. The primary benefit of sensors is their low price. High-performance and low-cost cams can now be mounted on the rear and front sides to provide a complete 360-degree view. Any changes to the road infrastructure do not affect visual detection and tracking. Optical sensors can track cars going around a curvature or at any time a track change more effectively. They also do not have the same problems with interference that active sensors do. One more important benefit of employing an optical sensor is providing a more detailed description of the vehicle. Passive sensor-based techniques are not as reliable as active sensors. The quality of collected images is greatly dependent on detection/classification, which can be easily influenced by lighting, fog, and other environmental factors. To obtain essential data from collected pictures, such devices want distinct and much-advanced image filtering algorithms and more excellent computational resources.

4.2 Passive Sensors

Passive sensors, such as acoustic and optical (camera) sensors, acquire data by receiving signals without releasing them. A sensing system based on audio cues or signal have been recently developed for accurate environment detection and monitoring of a vehicle approach. Using the gradient method, it started by creating a stable spatial feature from noisy acoustic observations. The spatial information is then filtered out using a particle filter and sequential state estimation. Actual world noise data obtained by vehicle-mounted microphones from outside cruise cars were used to test the proposed approach. The researchers provided a complete sound sensing control system experimental hardware for detecting automobiles by calculating road conditions using the feature of the urban road environment.

The traffic condition on the road was assessed by the server (a distant processor) based on these metric measurements. Motorcycles were also recognized for utilizing microphones in an array to the signal with a short wavelength component's uniqueness.

Visual data can be helpful as an overview of the cars in the area. Hence the utilization of optical sensor systems for collision mitigation approaches. Active sensors are more adept at anticipating vehicles than passive sensors. Detection can be done with a stereo camera, a single camera, or many cameras. The cameras can be installed on the inside of the windscreen to the backside mirror of the vehicle's body. Obtaining a full-degree view of the surrounding area may necessitate the use of many cameras in many circumstances. Because of their poor vision in low-light circumstances, infrared cams are used instead of regular cams for night-time detection.

For detection, single vision is used, whereas stereoscopic vision is used for three-dimensional positioning and tracking, which is characteristic of employing both monocular and stereovision cues. The cars are in the stereovision domain, monitored with extended Kalman filtering, utilizing a classifier in the monocular plane to determine the surface area of the earth and a disparity map to estimate the ground surface. It was also demonstrated how to employ track management to eliminate false alerts and improve accuracy. Using a similar approach, a set of boost detectors was developed for various automobile views, and candidate locations were by searching for spikes in the deviation range, it was confirmed. Monovision has the benefit of recognizing two things in 3D space that is close to one another and would lead to missing stereovision.

4.3 Sensor Fusion

Several sensor methods are often more likely to advance compared to a single sensor, resulting in more dependable and secure systems. As shown in Fig. 3, one sensor detects while the other confirms the obtained data or the first sensor detects, while the second verifies the data.

4.3.1 Radar and Vision

In recent years, the use of radar-vision integration for object tracking and perception has been gaining popularity. In this fusion, radar is primarily used to estimate ROIs or distances, while pattern recognition algorithms are used for recognition. Guardrail

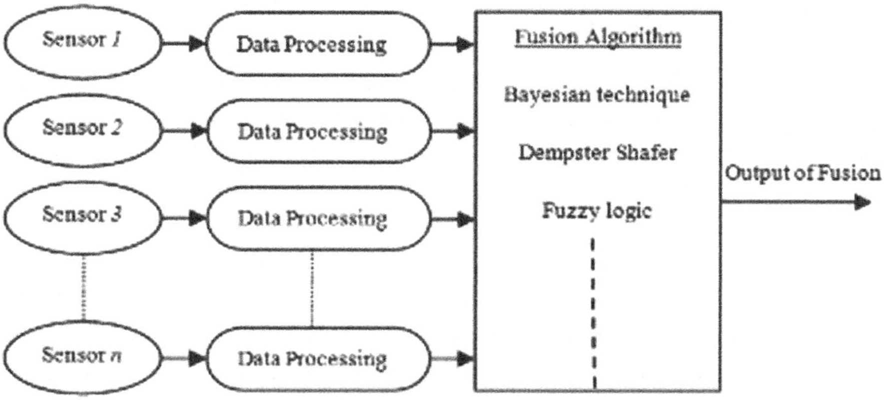

Fig. 3 Sensor fusion model

positions were chosen using radar images and the vertical regularity feature of the limited area in images (ROI) recognized vehicles. Automobiles were identified for their comparable use of symmetry, edge detection, and image filtering in pictures. Kalman filtration on radar data has been used to track, and distance detected objects. HOG, Haar, and Gabor features were used in classifier-based detection, and radar range finding was successful. The input image was assessed for salient spots using many visual properties such as orientation, intensity, colour, and motion. Once the vehicle was identified, the distance was computed using radar and visual data.

On the inverse projection projected image, the radar ranged the visual operations. In a global frame of reference, cars got monitored using Kalman filtration. For vehicle detection, a radar-vision online learning system was used.

4.3.2 Lidar and Vision

Researchers have looked at fusing lidar with monocular vision in recent years. Initially, detection and tracking were accomplished using lidar data to improve the accuracy of object identification, while obtained data from the lidar was accessed simultaneously for categorization.

The stereo vision system was used to confirm the presence of possible obstructions that multi-layer lidar had predicted.

Radar, vision, and lidar technologies were used to construct a multisensor method with extensively overlapped fields of view. On the sides of the car, there were two independent laser scanners and multiple overlapping short-range radar sensors, and three primary long-distance sensors were placed on the front.

4.3.3 Acoustic and Vision

The benefit of combining acoustic and video sensors is that they obtain complementary information for identification and monitoring. Acoustical data is analysed in the detection phase to analyse the target's arrival, which specifies the estimated target area in the video.

4.3.4 Radar and Lidar

A collection of radar and lidar data, obstacle characteristics connected with the prototype car's front bumper have been identified. Bayesian approaches were utilized to predict the state for improved detection and tracking, and tracking data from two separate systems were fused.

5 Vehicle Detection Schemes

In recent years, imaging technology has advanced dramatically, and Cameras are now more affordable, lighter, and of more excellent resolution than ever. Moreover, processing power has increased, and distributed processing platforms such as graphics processing units (GPUs) and multi-core processors have emerged. These equipment advancements make it possible to deploy computer vision algorithms for vehicle identification in the virtual environment as shown in Fig. 4.

The vehicle identification mode could be organized, with picture pre-processing removing noise and perhaps including contrast correction. The Cueing/Hypothesis

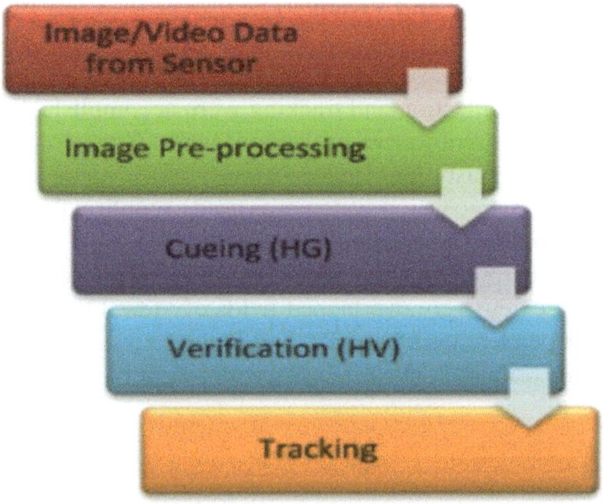

Fig. 4 Vehicle identification and monitoring system stages

Generation (HG) and Hypothesis Verification (HV) phases play essential roles. HV confirms the selected candidates as either vehicles or non-vehicles. HG attempts to determine probable object location in the obtained images and designate them as regions of interest (ROI). After it has been validated, A tracking function is used to track the object's motion in successive images. As the detecting process moves on from one phase to another, processing the number of data decreases. This enables more complex activities to work on a lower section of the picture. The following sections look at current cueing and verification methodologies, primarily based on vision-based object recognition algorithms.

5.1 Cueing

In radar and vision fusion systems, the region of interest is usually established by analyzing the range and information received by radar. The motion or visual aspects of the vehicle are used to identify ROIs in images captured by optical sensors. To identify objects at motion on their fluxes, motion approaches necessitate the analysis of numerous image frames. On the other hand, appearance-based vehicle detection examines a single image for visual signals to hypothesize vehicle placements. Stereo cueing approaches, on the other hand, can be either motion or appearance. The sections that follow clarify strategies to provide a review of the relevant literature.

5.1.1 Motion-Based Approach

To detect cars, motion-based algorithms use temporal information in sensor data. Optic flow patterns from a moving car are calculated by comparing local features or pixels across multiple images. Horn and Schunck proposed dense motion information whose intensity equaled that of all pixels in a picture. A set of unique features on a car, such as edges or colour splashes, were monitored using sparse motion information. Moving cars were segmented from the image after the optical flow fields were calculated and clustered based on their locations, magnitudes, and angles.

The optical flow fields were then created by tracking characteristics over numerous frames. The fields were clustered using a 'flow segmenter' upon their motion fluctuations. Finally, the clusters' frame and centroid were determined., and vehicles were proposed. The system requires a significant computational load to achieve real-time performance, so they implemented specific hardware acceleration.

5.1.2 Appearance Based Approach

Appearance-based approaches use certain aspects of a driver's viewpoint, such as shadows underneath the car, boundaries, symmetries, material, colouring, and vehicle lights, to extract ROI.

Feature of the Shadow

The shadow under the car is darker than the adjacent road surface. It serves as a cue for its presence. This can be seen in several ways, such as when a concrete road's spectrum was evaluated to find a limit for dividing darkening parts. Using dark areas in the pictures and boundary information, the position of cars was hypothesized. A cleverer strategy was devised, involving the use of shadows to approach cars. Road areas are first derived by delineating the image's centroid homogenous area, which was demarcated by edges. Then,

- Areas having intensity less than a defined level were designated as shadow zones.
- The horizontal corners of certain locations were then obtained based on the placements of these edges.
- Hypotheses were determined.

Edges Clue

The rear view of most automobiles shows both horizontal and vertical borders, which may be necessary for ROI creation. Researchers presented a multi-scale strategy to detecting a vehicle boundary using distinct detailed images, and the hint was used to determine. At the same time, an exciting region was found based on the cue using the shadow under the car to identify the location of a vehicle. A search that goes from rough to refined approach was developed in a coarser scan identified groupings of conspicuous corners. A refined search of these areas found objects with a rectangle shape. Motorcycles and bicycles are perhaps missed by the methods described above.

Corners

The general shape of a vehicle is a rectangle with four sides. This has been performed using the templates to evaluate automobile edges in the picture continuously. A probable vehicle was spotted if four sides were identical and had enough pixel values only at locations that coincided with the car's edges. A standard double-circle mask might be used to identify all corner types, speeding up the corner detection procedure. The types and locations of detected corners were then grouped. Finally, the characteristics of each cluster's corners were extracted and fed to determine if they belonged to the car, a Support Vector Machine (SVM) classifier was used.

Colour

Most vehicles have a similar colour scheme (body and warning light colour) contrast with the surface of the road and surrounding vehicles. This was used to divide automobiles from photos by defining the vital vehicle colour with a colour transform model. The combination of red warning lights and yellow signaling lights, for example, warning lights, on the other hand, is considered an indication for vehicle recognition using colour segmentation algorithms.

Texture

The texture of vehicles differs from their surroundings, including the roadway and greenery. Statistical studies on entropy were utilized to break out ROIs in photographs.

Headlights/Taillights

In complete darkness or darkened environments, the performance of a standard vision sensor is severely harmed. The information in image/video data is distorted by light reflection induced by vehicle motion. As a result, most of the strategies mentioned in this study are unreliable for nocturnal vehicle detection. Because visual clues (e.g., shadows, edges, tire colour) fail to deliver in night-time contexts, they are challenging to navigate. Vehicle headlights and taillights are the only features that may be used to hypothesize and detect automobiles at night.

5.1.3 Verification Stage

There are two types of verification techniques:

(1) Template matching and
(2) Object classifier.

Template Matching

This method employs a present template to calculate a measure of closeness by measuring region of interest correlation with the templates. Because there are so many distinct models of automobiles with so many varied looks, a 'loose' and generic template with everyday aspects of a vehicle is employed. So, the size of a vehicle changes depending on its range from the cam. The connection test is usually performed on multiple ROI scales. To get a consistent result, the image intensity is also normalized before the correlation test.

Adaptive templates were suggested, where a pattern for further identification is produced automatically by resizing the picture of the identified vehicle after it has been identified using a generic template. The robustness of the pattern matching method was increased by updating the templates, which has been evaluated using the targeted edges, surface.

Classifier-Based VehicleVerification

This method employs two-class image classifiers to distinguish the vehicle from non-vehicle options. From training pictures, a classifier analyses the visual aspects of a vehicle. This program uses a guided learning method with a large number of tagged vehicle and non-vehicle images. Artificial Neural Networks (ANN) and Support Vector Machine (SVM) are most used classification systems for vehicle verification. Training photos are initially pre-processed to extract descriptive information to aid categorization. It is critical to choose characteristics carefully to attain that good categorization results can be achieved. A solid selection of features will be able to catch most of the differences in the look of a vehicle.

6 Tracking

To increase vehicle recognition accuracy and speed, tracking takes the benefit of information stability. It has two primary functions: estimating vehicle motion and position and maintaining track of previously recognized vehicles that are not identified in a particular image when sorting out false alarms. Measuring and sensing ambiguity are two issues that tracking may overcome, administration of data correlation and tracking. Typical hypothesis development procedures are made up to maintain performance while monitoring production reductions. On the other hand, monocular tracking methods often verify dynamics in pixel, whereas stereo-vision approaches verify dynamics in meters. Vehicle monitoring methodologies such as Kalman, Extended Kalman, and Particle filters using various sensors are routinely applied to monitor and identify the motion of observed vehicles in image planes using traditional tracking and filtering techniques.

7 Collision Avoidance Challenges

A significant problem in implementing a successful collision avoidance strategy is to design the vehicle detecting systems with optimum accuracy and robustness in real-time. Significant efforts have been made in this scientific area, and various methodologies and systems have already been proposed. Even though several experimental cars were evaluated to demonstrate the effectiveness of developed models, a thoroughly reliable, durable, and authentic system is still to be unveiled. Google's self-driving car had taken a giant stride ahead in creating self-cars equipped with advanced sensors and computer vision systems. In constructing a real-world CAS suitable for urban streets, traffic congestion, bikes, cyclists, junctions, people, road signs, and others all add additional barriers and technological challenges. The effectiveness of a CAS will be measured by the number of correct detections vs. the false alerts.

7.1 Sensor Challenges

Choosing the correct sensors is the earliest crucial step in building a functional and resilient CAS. Among the specific targets are enhanced spectral sensitivity, pixel density, image quality, and computing capabilities. Active sensors work well in various weather conditions, including at night, and their prices are reasonable, except for the 3Dlidar scanner. CAS with this scanner is undoubtedly too expensive, costing almost as much as the car alone. Conventional camera in the industry lack with the image quality required to operate in traffic under adverse conditions.

Night vision cameras are required to allow daylight and evening road activities without blurring. These cameras switch to Infrared (IR) mode whenever the lighting intensity falls below a predetermined level. Nevertheless, some of those sensors may have limitations, such as a narrow field of vision. High-resolution camera available at a reasonable cost provide significant benefits. On the other hand, high resolution necessitates the processing of more data (pixels), which increases processing time and necessitates the use of powerful computer resources.

To deal with busy and complex traffic conditions, vision-based systems and algorithms must progress into more sophisticated solutions. Multiple sensor fusion could result in significant increases in CAS performance and a higher level of security and reliability. With a multisensory approach, the system may gather more extensive and precise information about the environment that is impossible to detect with a single sensor.

8 Conclusion

On-road vehicle detection approaches for Collision avoidance systems, particularly automatic vehicle identification, have gotten much attention in recent years. However, because of faulty collision avoidance systems and diverse on-road scenarios, the difficulty remains. It begins with a sensor performance comparison, which reveals that active sensors operate well in various weather conditions but suffer from interference concerns. This review paper recommends a combination of sensors (active and passive), cueing, and verification based on classifier as the best option if you are looking for a lower cost, dependable solution for collision avoidance system design for the automotive industry. For future Collision avoidance system design, these points must be given extra attention and worked out.

References

1. Yue, L., Abdel-Aty, M., Wu, Y., Wanga, L.: Assessment of the safety benefits of vehicles advanced driver assistance, connectivity and low level automation systems. Accid. Anal. Prevent. **117,** 55–64 (2018)
2. Bellaa, F., Russoa, R.: A collision warning system for rear-end collision: a driving simulator study. Roma TRE University - via Vito Volterra n. 62 - 00146 Rome, Italy. Proc. Soc. Behav. Sci. **20**, 676–686 (2011)
3. Crash data collected by the U.S. National Highway Traffic Safety Administration (NHTSA)
4. Yadav, R., Dahiya, P.K., Mishra, R.: Comparative analysis of automotive radar sensor for collision detection and warning system. Int. J. Inf. Technol. **12**(1), 289–294 (2020)
5. Shen, Q.: Design of backward collision warning and avoidance system when on-street parking using LiDAR. https://doi.org/10.1109/CDS52072.2021.00048
6. Al-Smadi, A.M., Al-Ksasbeh, W., Ababneh, M.: Intelligent automobile collision avoidance and safety system. In: 2020 17th International Multi-Conference on Systems, Signals & Devices (SSD '20)
7. Jansson, J.: Collision avoidance theory with application to automotive collision mitigation collision avoidance theory with application to automotive collision mitigation. Linköping Studies in Science and Technology. Dissertations No. 950, Department of Electrical Engineering Link¨oping University, SE–581 83 Linköping, Sweden Linköping 2005. ISBN 91-85299-45-6. ISSN 0345-7524
8. Mukhtar, A., Tang, L.X.T.B.: Vehicle detection techniques for collision avoidance systems: a review. IEEE Trans. Intell. Transport. Syst. **16**(5) (2015). https://doi.org/10.1109/Tits.2015.2409109
9. Kashevnik, A., Shchedrin, R., Kaiser, C., Stocker, A.: Driver distraction detection methods: literature review and framework. https://doi.org/10.1109/ACCESS.2021.3073599
10. Beanland, V., Fitzharris, M., Young, K.L., Lenné, M.G.: Driver inattention and driver distraction in serious casualty crashes: data from the Australian national crash in-depth study. V. Beanland et al., Acci. Anal. Prevent. **54**, 99–107 (2013)
11. ANCAP: South-East Asian crash test program commences, ed. (2012)
12. Gietelink, O.: Design and validation of advanced driver assistance systems. Ph.D. dissertation, Technical Univ. Delft, Delft, The Netherlands, 2007, ISBN: 978-90-5584-091-5
13. Zhang, X., Liniger, A., Borrelli, F.: Optimization-based collision avoidance. liniger@control.ee.ethz.ch, June 12, 2018
14. Zhao, Z., Zhou, L., Zhu, Q., Luo, Y., Li, K.: A review of essential technologies for collision avoidance assistance systems. Adv. Mech. Eng. 2017 **9**(10), 1–15. https://doi.org/10.1177/1687814017725246
15. Doi, A., Butsuen, T., Niibe, T., Takagi, T., Yamamoto, Y., Seni, H.: Development of a rear-end collision avoidance system with automatic brake control. JSAE Rev. **15**, 335–340. Society of Automotive Engineers of Japan, Inc. and Elsevier Science B.V. (1994)
16. Dahl, J., de Campos, G.R., Olsson, C., Fredriksson, J.: Collision avoidance: a literature review on threat-assessment techniques. IEEE Trans. Intell. Vehicles **4**(1) (2019)
17. Hamid, U.Z.A., Pushkin, K., Zamzuri, H., Gueraiche, D., Rahman, M.A.A.: Current collision mitigation technologies for advanced driver assistance systems—a survey. PERINTIS eJ. **6**(2), 78–90 (2016)
18. Nmngani, A., Akyurt, M.: Review of vehicle collision avoidance systems. In: The 6th Saudi Engineering Conference, KFUPM, Dhahran, December 2002, vol. 3, pp. 413–428
19. Manjunath, K.G., Jaisankar, N.: A survey on rear end collision avoidance system for automobiles. Int. J. Eng. Technol. (IJET) **5**(2) (2013). ISSN: 0975-4024
20. Sreevishakh, K.P., Dhanure, S.P.: A review paper on automotive crash prediction and notification technologies. 978-1-4799-6892-3/15 $31.00 © 2015 IEEE. https://doi.org/10.1109/ICCUBEA.2015.197

21. Vaishali, B., Jeyapriya, A.: A survey on intelligent system for automobile collision avoidance at intersection of roads. In: International Conference on Innovations in Information, Embedded and Communication Systems (ICIIECS) (2017) 978-1-5090-3294-5/17
22. Adla, R., Al-Holou, N., Murad, M., Bazzi, Y.A.: Automotive collision avoidance methodologies sensor-based and ITS-based. 978-1-4799-0792-2 IEEE
23. Cherng, S., Fang, C.-Y., Chen, C.-P., Chen, S.-W.: Critical motion detection of nearby moving vehicles in a vision-based driver-assistance system. IEEE Trans. Intell. Transport. Syst. **10**(1) (2009)
24. Mushtaq, M.I., Kansal, M.: A review on vehicle collision avoidance system in networks. Int. Res. J. Eng. Technol. **3**(5), 2747–2749
25. Madhu, B.K., Kota, K., Surabhi, K., Nikhil, U., Yashwanth, M.: Vehicle collision avoidance system. Int. Res. J. Eng. Technol. **7**(6), 1206–1210
26. Kathirvelu, M., Shree, A.C.N., Manasa, M., Naveen, L., Karthi, N.: Vehicle speed control and accident prevention system in smog zone, 1916 (2021) 012093, ICCCEBS 2021. J. Phys. Conf. Ser. https://doi.org/10.1088/1742-6596/1916/1/012093
27. Haselhoff, A., Kummert, A.: An evolutionary optimized vehicle tracker in collaboration with a detection system. Proc. 12th Int. IEEE Conf. ITSC 1–6 (2009)
28. Stutts, J.C., Reinfurt, D.W., Staplin, L., Rodgman, E.A.: The role of driver distraction in traffic crashes. AAA Found. Traffic Safety, Washington, DC, USA (2001)
29. Driver fatigue and road accidents: a literature review and position paper. R. Soc. Prevent. Accid., Birmingham, UK, pp. 1–24 (2001)
30. Nilsson, J., Ödblom, A.C.E., Fredriksson, J.: Worst-case analysis of automotive collision avoidance systems. IEEE Trans. Vehicular Technol. **65**(4) (2016). https://doi.org/10.1109/Tvt.2015.2419196
31. Biassoni, F., Ruscio, D., Ciceri, R.: Limitations and automation. The role of information about device-specific features in adas acceptability. F. Biassoni et al., Safety Sci. **85**, 179–186 (2016)
32. Coelingh, E., Eidehall, A., Bengtsson, M.: Collision warning with full auto brake and pedestrian detection—a practical example of automatic emergency braking. In: 2010 13th International IEEE Annual Conference on Intelligent Transportation Systems Madeira Island, Portugal, September 19–22 (2010)
33. García-Costa, C., Egea-López, E., García-Haro, J.: A stochastic model for design and evaluation of chain collision avoidance applications. C. García-Costa et al. Transport. Res. Part C **30**, 126–142 (2013)
34. Grime, G., Jones: Car collisions—the movement of cars and their occupants in accidents. Proc. Inst. Mech. Eng. Automob. Div. **184**(1), 87–136.

Motion Planning and Manoeuvring in Self Driving Cars

Joseph Nixon Kiro, Srishti Anand, Shalini Mahato, and Laxmi Kumari Pathak

Abstract Lane keeping, path planning and decision making are the most important features of self-driving cars. Cooperative adaptive cruise control and platooning are notable applications in autonomous cars. It expands the use of street foundation assets through composed control and platooning of individual vehicles with the capability of improving both traffic wellbeing and productivity. In this study, end-to-end lane keeping and path planning are explored. Self-driving cars are presently recognised as the core of industry research because of its multi-faceted quality vehicle benefits which includes advanced security, reduced blocking, minor manufacturing and improved mobility. When transporting people or commodities from one location to another, motion planning methods include locating a track to follow, avoiding roadblocks and designing the optimum route that ensures security, ease and effectiveness. The purpose of this study is to review present approaches and evaluate the different techniques for the motion planning of self-directed on-road driving that consists of locating a lane, finding for the safest manoeuvre and shaping the main possible route. The study demonstrates a critical evaluation of each of these processes in terms of their benefits and drawbacks.

1 Introduction

Cooperative driving depends on remote communications among vehicles and side of the road foundation, along these lines giving the likelihood to trade data past the view of person vehicles, and to acquire data that can't be recovered by means of on-board sensors. Thus, this prepares to make self-coordinating conduct inside and between gatherings of vehicles, focusing on expanded traffic stream and traffic wellbeing, while diminishing fuel utilization and outflows, specifically when joined with mechanization of the individual vehicle movement. In view of this potential, helpful driving right now gets plentiful consideration on a worldwide scale [1]. Automated

J. N. Kiro · S. Anand · S. Mahato (✉) · L. K. Pathak
Amity Institute of Information Technology, Amity University Jharkhand, Ranchi, Jharkhand 834002, India

A. E. Hassanien et al. (eds.), *Virtual and Augmented Reality for Automobile Industry: Innovation Vision and Applications*, Studies in Systems, Decision and Control 412, https://doi.org/10.1007/978-3-030-94102-4_2

driving innovation is at present perceived as one of the potential arrangements that may can possibly further develop street traffic security and proficiency.

Automated vehicles as a rule use locally available sensors like LiDAR, Radar and additionally camera to detect traffic ecological data for way arranging and afterward to direct vehicle movement control to follow the arranged way [2]. Self-driving cars uses Convolutional Neural Networks (CNNs) model which catch raw picture structures as information and yields the steering points in like manner. After that the most significant job for an autonomous vehicle is towards track down the legitimate controlling the vehicle contribution to keep up with it in path. The path markings are typically distinguished by some picture handling strategies, for example, shading improvement, hough change, edge location, and so on [3]. The main purpose of the study is to analyse how self-driving car chooses it's path, eliminate it's obstacles and find the best and optimal way to reach the destination while following all the traffic rules. All these are planned in a memory and further executed in algorithms which it received in the form of digital map. These maps must be reliable, adaptable planning, robust. These are the essentials when it comes to urban mixed traffic scenario.

Urban Challenge and Grand Challenge were organized by the DARPA (Defense Advanced Research Projects Agency). The DARPA settled the Grand Challenge for the initial time in 2004 where unmanned automobiles run the desert followed by the second Grand Challenge in 2005 and the Urban Challenge in 2007 because of the advancement in the field of automotive sector, automobile are capable of independent driving in the actual world, while maintaining every rules and regulation of traffic signals [4].

The purpose of this study is to review present approaches and evaluate the different techniques for the motion planning of self-directed on-road driving that consists of locating a lane, finding for the safest maneuver and shaping the main possible route. The study demonstrates a critical evaluation of each of these processes in terms of their benefits and drawbacks which would help the researcher to develop technology which overcomes the limitation of self driving car.

The paper is structured as follows. In Sect. 2, Architecture i.e. mechanism of self-driving cars has been discussed. It includes motion planning, search space planning, planning techniques, maneuver planning, crash detection and intersection passing situation. Section 3, Limitations discusses the drawbacks as well as solution for overcoming them has been discussed. Section 4, Test Results highlights test results of various competition and challenges. It also includes the results of various studies done in the area of self driving cars. A brief Discussion has been done in Sect. 5. Section 6 includes the Conclusion of the finding of the study.

2 Architecture

The architecture of the arrangement of automated vehicles is regularly coordinated keenly based on two principle division: the Decision-making system and the Perception system. The perception system is substantially separated into numerous subsystems liable for certain assignments for example, static obstacles mapping, moving obstacles detection and tracking, road mapping, autonomous car localization, traffic signalization detection and recognition. Whereas the decision-making system is normally separated into numerous subsystems to manage certain works, like behaviour selection, path planning, route planning, motion planning, obstacle avoidance plus control.

The Perception system manages to assess the State of the vehicle and for making an inside portrayal of the surroundings, utilizing information captured by on-board sensors. The decision-making system navigates the vehicle from its primary to the finishing destination characterized by the consumer allowing for the present vehicle's situation and the inner portrayal of the environment, just as traffic rules and traveller's wellbeing and solace. The Localizer subsystem assess the vehicle's State (position, linear and angular velocities and so forth) in connection to static maps of the surroundings. The Moving Objects Tracker subsystem (MOT) obtains the Offline Maps and the vehicle's State, and determines and tracks, i.e., computes the position and speed of the closest moving hindrances. The Traffic Signalization Detector subsystem (TSD) detects the traffic signals, whereas the Path Planner subsystem evaluates the surroundings of the car just as traffic rules. The Behavior Selector subsystem manages the current driving behavior, for example, path keeping, crossing point, dealing with traffic signal etc. The Motion Planner subsystem computes a direction from the current self-driving car's State to the current objective, that sticks to the Path characterized by the Behavior Selector fulfils vehicle's kinematic and dynamic limitations and gives solace to the travellers. The Obstacle Avoider subsystem gets the Trajectory processed by the Motion Planner and it makes the necessary changes like decreasing the speed, to keep away from crashes. At last, the Controller subsystem receives the Motion Planner direction, ultimately altered by the Obstacle Avoider subsystem, and registers and sends Effort commands to the mechanism of the steering wheel, accelerator and brakes to cause the vehicle to carry out the Modified Trajectory. Figure 1 shows the mechanization framework architecture of self-driving cars [5, 6].

2.1 Motion Planning

In a cooperative driving system, automated vehicles can not just drive independently and securely utilizing locally available sensors, yet additionally have the capacity of platooning with adjoining automated vehicles through collaboration utilizing vehicle to vehicle (V2V). Cooperative controllers for automated vehicles in platoon comprise of three useful parts:

i. Preparing the joined data including vehicle movement states furthermore, traffic climate through locally available sensors and V2V.
ii. Producing control targets and imperatives for a vehicle move progress rule dependent on hybrid automata model under the current traffic occasions.

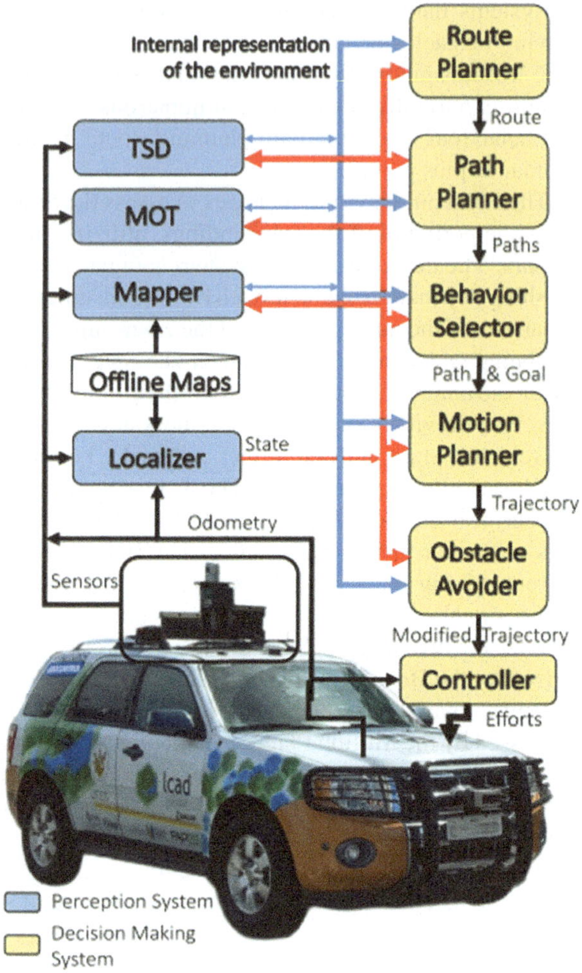

Fig. 1 Architecture of the mechanization framework of self-driving cars [5]

iii. Yielding control orders, for example steering angles and longitudinal increase in speed to achieve a vehicle's parallel and longitudinal movement control [2].

Autonomous vehicles have shown it's true potential and progress in the recent technology and higher developed driver assistant systems and are predict to be the sustainable future for improved road safety, capable traffic flow and decreased fuel expenditure, while improving, mobility and hence general welfare [7].

The motion planning input (sensor input) and output hang on module like obstacle tracking, computationally intensive routine data fusion and control module run in parallel with memory consuming in planning. The necessary part of planning is reliable, adaptable and robust. These three play the significant role in urban cities where diverse traffic picture occurs. The senor structure and complement these input with data from digital road map for complete workstation planning [6, 8].

In wide-ranging, planning for automobile driving is separated into different modules as follows.

2.1.1 Path Representation

The information is gathered from comma.ai driving dataset and is stated that the dataset has 7.25 h of driving data, 11 video clips recorded at 20 Hz and other calculations like steering angle, velocity, GPS data, etc. The size of the image frames is 320 × 160 pixels, cropped from original video frames [3]. A path is represented as n cubic equations and the points are separated into different segments. A beginning stage of every path segments is situated to the origin, after that an coordinate is changed to make a slant to get zero, accordingly performing bend into Eq. (1).

$$y_i = a_i x_{path^3} + b_i x_{path^2} \tag{1}$$

$$p_i = [x_{i,ENU}, y_{i,ENU}, h_{i,ENU}, a_i, b_i, l_i, v_i]^T \tag{2}$$

A *i*th path can be expressed simply by an array in (2), which decreases the size of the array altogether contrasted with characterizing it by way focuses and can be utilized for search of presence of impact or new way age advantageously. Here xi, ENU, yi, ENU and hi, ENU are the origin point and azimuth of path while Li and Vi are the lengths of path and maximumspeed (limit speed) of relating path [9].

2.1.2 Path Planning

An automated vehicle can keep away from deterrents by following the produced direction utilizing the Artificial Potential Field (APF) approach. Here, an all around planned state of the potential field addressing obstructions ahead assumes an essential part for the hindrance aversion adequacy. The repulsive field consistently has an impact area where the more limited the distance to the objective deterrent, the more noteworthy the repulsive force is. Additionally, outside the impact locale, the horrendous power will quickly lessen to the base or zero [2]. The main purpose of path planning is to reach its destination with no clash with obstacles with geometric trace. Path-planning basically deal with finding a geometric trace from an primary configuration to a set finally configuration such that each arrangement and state on the passageway is a realistic [10, 11].

2.1.3 Route Planning

It specially deals with finding the best possible route from initial point to destination point keep in mind with real-time traffic information [10]. This examination estimates no slip among wheels and the surface, and utilizes a model which expects to be the right and left wheels as one to control sidelong bearing of a vehicle [7, 9].

2.1.4 Maneuver Choice

It refers to high- level decision making operation like 'overtaking', 'going straight' or taking turn 'turning left and right' according to the need. But this high level performance is taken into consideration only if the vehicles are performing safely with respect to traffic or other rules. Path planning in other term refers to dealing with high level decision in the best form choose for the vehicles, according to the path planning [7, 10].

2.1.5 Trajectory Planning

It is also known as direction planning, trajectory planning mainly focus on the real-time arrangement of vehicle's shift from viable state from another, fulfilling vehicle's kinematics limits depending upon automobile dynamics and controlled by the navigation console, taking consideration of lane boundaries, obstacles like ditches, rough ground etc., traffic rules. Trajectory planning algorithm follows two ladder:

i. First step include low resolution search space
ii. Second step include high resolution search space.

Automobile move on the road, it travels on the road by taking look on the digital map for information and sensor. The digital map are use for representing every information like road direction and turning, road network [10].

2.2 Search Space Planning

Planning a trip for an automobile on the road, which inquire for the path represent in surrounding i.e., physical path. So physical path must be present in configuration state or state space. The state space consists of representations like position, linear or angular velocities, orientation. As automobile travels on the path, by taking consideration of digital map for many information and it's also important for dealing with search space, search planning use five graph techniques [12].

i. Voronoi diagram
ii. Occupancy grid
iii. Costmap
iv. State lattices
v. Driving corridor.

Figure 2a represent voronoi diagrams, it create a way which maximize the space between the medium and the obstacles. It give the complete view if the way found in free space. But it is use for limited to static environment and it provide the discontinuous edge [13].

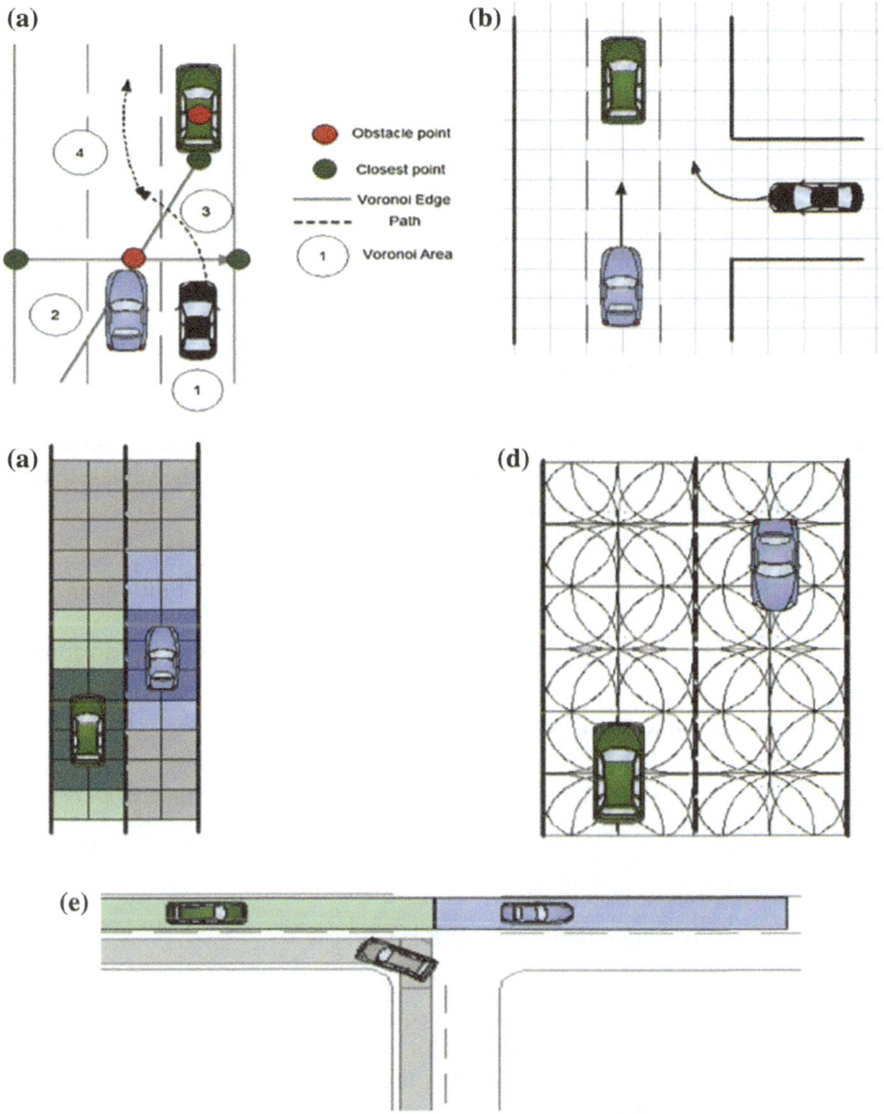

Fig. 2 Graphs used in planning. **a** Voronoi diagram, **b** occupancy grid, **c** costmap, **d** state lattice, and **e** driving corridor [9]

Figure 2b, c represent occupancy grids cost map method, it is has grid which means cell like structure with probability that each of the cell in occupied by the obstacle with low computation cost. But it also face the program with the different vehicle dynamics and it's also seen error in finding the obstacles [13].

Figure 2d represent state lattices, it's generalize grids which is made up primitive paths of square or rectangle in repetitive in continuous form. And these paths are connect for the automobile. But it also has limitation, the main problems with curvature with grid, restricts motion of vehicles and this method also face difficulties in dealing with evasive manoeuvres [13].

Figure 2e represent driving corridors, it's basically depend on the lane boundary and the other obstacles given on complete digital maps made by (SLAM) simultaneous localization and mapping. It provide continuous collision which initiate free space for the movement of vehicle. Higher computation cost and limitation on motion are the drawback of this method [13].

2.3 Planning Techniques

This part is all about how we use planning techniques is being used in on-road driving, how we use the existing studies for finding best possible path for the vehicle to drive the vehicle on road without colliding to the obstacles.

These types planning include mainly three level of planning, and those are:

i. Searching for the best geometric path for the automobile to follow
ii. Searching best maneuver to carry out.
iii. Take decision the finest trajectory to follow for achieving the optimization.

2.3.1 Incremental Search

Basically it is a technique which is used to find the best possible path from re-using the information from the previous search history for maximize the speed of search [14].

The two technique which in use are.

i. Rapidly—exploring random trees (RRTs)
ii. Lattice Planners (LP).

Lattice planner and RRTs both use data structure (lattices and trees respectively) RRTs use kinematics and real-time feasibility for quick searching of free space. In RRTs it use sophisticated decision skill are applied for checking the collision. Whereas lattices have less computational power is needed and lattices are appropriate for dynamic environment. Easy exploration is available in both scenario for choosing the path are supplied to the planning unit for the automobile to follow.

2.3.2 Local Search

This search focus on finding possible path in limited horizon because searching in entire graph is complex. Due to this it's most popular technique used for self-car driving [15]. Figure 3 shows the path planning and Table 1 shows the comaparision beween action and state space search.

Generally, path is generated in two form:

i. Lateral shifts in the action space of the vehicle and
ii. Lateral shifts in the state space of the vehicle (Fig. 3 and Table 1).

2.4 Maneuver Planning

Maneuver planning, includes techniques which predict the behavior of mutually the motorized and non-motorized traffic applicants and assesses the adjacent traffic circumstances, accordingly arming the driverless vehicle to settle on its best manoeuvre. Methods which are explained in this piece work on a more high-level root. Manoeuvre planning shifts away from exploring a trail or creating a route; in its place performing as a 'brain' which filters the outcome of path look for, cooperates with other traffic participants plus gives the approval for the geometric path before it is changed to a reasonable trajectory.

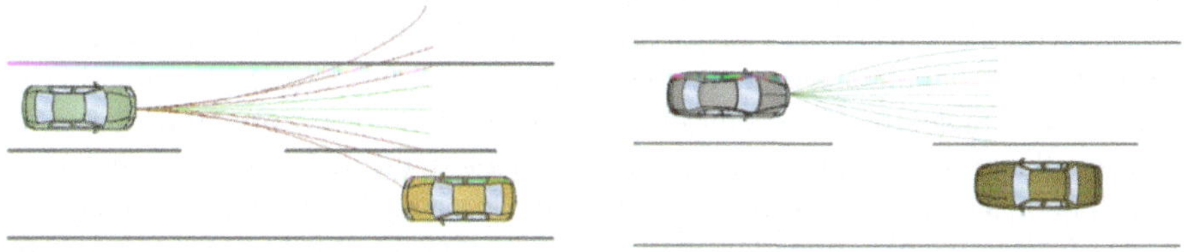

Fig. 3 Path planning [9]

Table 1 Comaparision beween action and state space search [10]

Action space	State space
Pros i. Since the eventual states are examined from the vehicle's kinematic and dynamic states the paths are realistic	Pros i. Paths are very realistic because of the productive examination of search space ii. Search space are very productive and eloquent iii. It has the capacity to undertake the global guidance and environmental shape
Cons i. It is very challenging to handle the state-space reaction and the intense activities ii. A high-dimensional action space is computationally very expensive	Cons i. It is very difficult while overhauling, since it is not stable while on curves

Motion planning methodology concludes simulation of multiple case study. The roads may be straight with some arc and with few curvature or banked at some angle. In a typical situation, most of the vehicles which will stick to traffic rules. In highway vehicle obstacle prediction is narrowed to unidirectional movement, lane changing, braking, maintaining speed with appropriate distance with vehicle, accelerate, decelerate [16].

2.4.1 Position Maintenance

The vehicle must maintain the front and back space for safety measures. The anterior judgement is of sustaining the constant speed and the lateral judgement is to hold the vehicle inside the lane margins. The vehicle must follow the consecutive forward car with an appropriate space between as precautionary measures [16].

2.4.2 Alteration of Path During Overtaking

The decision of alteration of path is taken under obstacle hinderance. For maintaining a constant flow of vehicle, shift to left or right according to the requirements. Hence, the motion planner must satisfy the space and speed is adequate to continue in the following lane or shift to adjacent lanes to provide the vehicle a safe space.

Lane Change Maneuver (LCM) is performed with the help of a virtual direction that isn't straightforwardly estimated. Bezier curve is one of the regular curve generation algorithm because it very easy to manage the curvature of the path generated by setting the control points. During the path change, thinking about the current driving pace of the vehicle, the horizontal increase in speed is powerfully produced in the direction planning stage progressively. In this method, the curvature of the produced Bezier curve is assessed to restrict the lateral movement. The speed is then increased by the LCM path generation algorithm and it changes the lane by first affirming for nearby path. After that, the direction planning block produces potential ways, and the generated local path is assessed to determine whether it meets the given speed increase requirement. If it meets the required acceleration, then it selects a path [17]. Figure 4 shows block daigram representation of lane change maneuver using Bezier Curve Algorithm [18].

The vehicle indicates through light for keep going on a lane while the adjacent lane vehicle follows it and then shifts the lane and shift back while overcoming obstacle. This requires constant lane changing then overtake a vehicle or constraint and another lane shift to move to precious lane.

2.4.3 Road Merge

It is to enter one ongoing traffic flow in highway or to exit out from the existing traffic flow. In this condition, vehicle should alter its speed and maintain safe space from

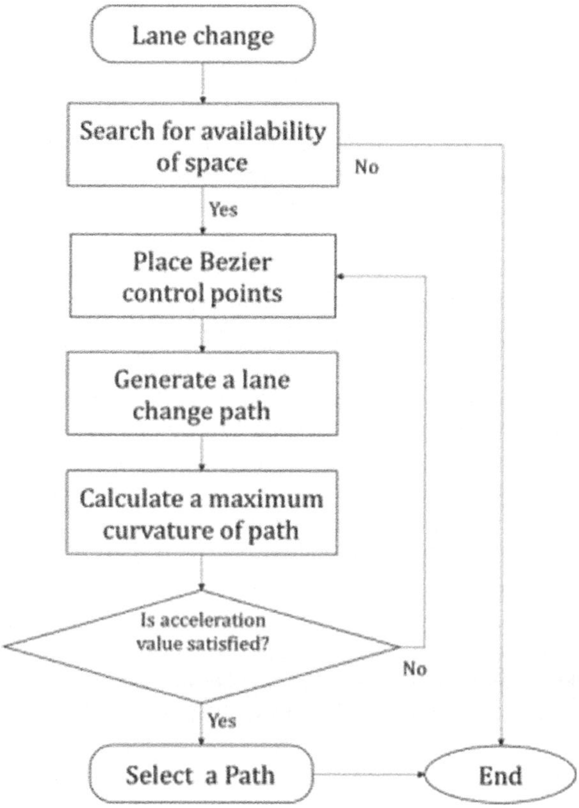

Fig. 4 Block diagram of the lane change maneuver using Bezier curve algorithm [18]

other vehicles. To figure out the path reduction the whole situation is first improved by separating the two platoons into modules comprising of three vehicles. These matched trios are then searched into the two units and cooperate for consolidating and making distance among them. Then, at that point, inside every one of these trios, a comparative association convention is carried out. The choice on who will be in every trio, is for the most part dependent on the general situation of the vehicles. That is, a vehicle most presumably likes to converge in front of a vehicle which is the nearest, for certain special cases such as converging in the middle of two trucks. The merging technique depends upon the input control through cooperative adaptive cruise control (CACC) with the vehicle just in front of it as an object and the obstacle avoidance (OA) regarding the combined vehicle in the nearby path. This procedure empowers all vehicles to choose for their fitting sets locally, without a need of a lead platoon to do the task [1, 17].

2.4.4 Toll Lane

It is to decide whether to merge and slow down for a toll lane or to go on to the traffic flow.

The main difference in highway and city driving is that the traffic flow in the highway is mostly unidirectional and a lesser shift in lanes whether in city driving lesser space is maintained in between vehicle and the traffic flow is of multidirectional nature. The vehicle mechanics in highway is of straight forward nature with lesser acceleration-deceleration, lower braking and a constant speed maintenance, less lane shift and a more risk due to high speed with a overall stable drive. The algorithm which will comply all these situation will be approved for a practical exercise in highways.

Even after following safety protocols, algorithms and high computation ability, we have to specify some highway constraints. Constraints is of two types specifically hard constraints and soft constraints. Under hard constraints lies protocols which are absolute and cannot be relaxed for autonomous vehicle including environment safety, safe driving norms, vehicle collision. On the contrary, soft constraints can be relaxed to a limit that includes ride customization, speed control, energy efficiency and comfort. Other constraints depend on vehicular kinematics, vehicle being a 3D object can only move freely in a 2D space, path and mechanics of movement should be differentiable and under limitation of autonomous movability of vehicle. This kinematics make complex algorithms with an increase in degree of freedom the complexity of model increases. Degree of freedom being one of the basic parameters of motion planning with architecture should be marked for a proper motion. Evolution of vehicle according to time being the last problem faced. The motion planned should be usable, safe, develop with time and efficient with progress [17].

2.5 Crash Detection

Crash detection algorithm is separated into two sections: 'Detection from vehicle position to the control point' which decides the presence of obstructions over the direction produced concurring to the revolution span of a vehicle between current vehicle position and the control point. And, 'Detection from the control point to the impression region' which decides presence of impediments by expanding the way to one side and left side to make an impact discovery limit dependent with the understanding that vehicle follows the way accurately [8]. Figure 5 shows representation of the collision avoidance system [9]

Here, P_k is the original path, $P_{k+1,1}$ and $P_{k+1,2}$ are the diversion points to avoid the collision. $P_{k+2,1}$ and $P_{k+2,2}$ are the convergence points to the path. P_{k+3} is the original path after avoiding the collision.

The information's such as an online map that addresses the surroundings around the vehicle, the present status of the vehicle in the online map, and the path planned by the motion planner subsystem are sent to the obstacle avoidance subsystem at

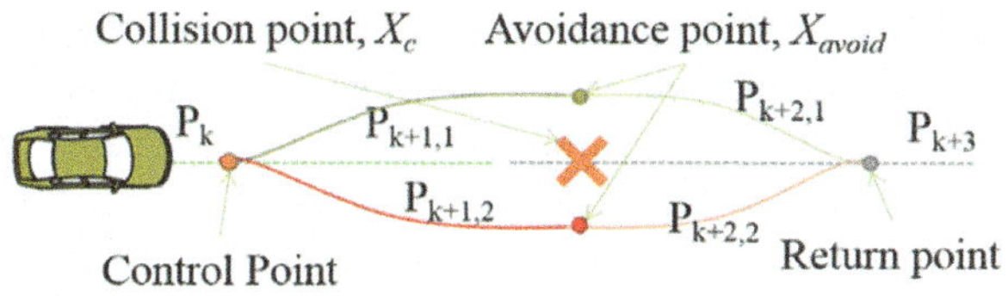

Fig. 5 Diagrammatic representation of collision avoidance system [9]

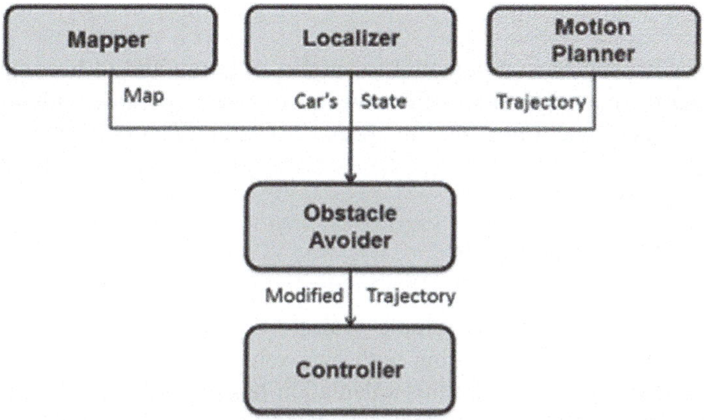

Fig. 6 Block diagram of obstacle avoider system [5]

each point, and thereafter, the obstacle avoider subsystem recreates the path and if an accident is anticipated to happen in the same direction, it reduces the speed of the car to avoid the collision [5]. Figure 6 shows the block daigram representation of obstacle avoider system.

2.6 *Intersection Passing Situation*

To figure out the crossway passing situation the vehicle has to have the option to execute the convergence crossing situation, a supposed Competition Zone (CZ) is characterized, being a round area with its middle concurring with the focal point of the *T* crossing point, in which vehicles are permitted to respond to each other. At the point when all vehicles associated with the situation enter the CZ simultaneously, the supposed target vehicle assignment (TVA) subsystem is initiated. This checks the path, destination and need of every vehicle to shape a virtual platoon. Every vehicle computes a virtual inter-vehicle distance between the relegated target vehicle and itself. Then, at that point, this distance is taken care of to the vehicle-following controller to understand an ideal virtual between vehicle distance. At the crossway, the vehicle sends the intersection's computer through wireless communication to tell it that a vehicle is coming. And afterwards the wireless communication again holds its entrance, space, and time in the crossing point. The computer in a split second evaluates every vehicle's position, speed, and size comparative with different vehicles coming from all directions and simulates all possible paths of the vehicle. [1].

3 Limitations

Autonomous driving have a potential to sustain in future and to become feasible. Even though there are limitations for now and still incomparable with human skill of driving. Some of the limitations that rests with autonomous vehicle are:

3.1 Hindrance Operation

It includes predicting other vehicle trajectories with velocity, acceleration-deceleration, brakes and comparing with own vehicle trajectories. This requires a lot of computational power and this calculation has to be down at every ongoing movement. These trajectories are not calculated for whole traffic flow and hence results in a few minor clashes. A wide approach did not considered hinderance in motion that being one of the major drawbacks. There is also a skill, level of understanding deficit when autonomous vehicle is compared to human driven vehicle. There being a whole gap and both potential for development till it is usable in real world [4, 19].

3.2 Vehicular Motion

In autonomous vehicle, most of the models are based on two wheeler vehicle and not a typical car model and is unable to operate the full ability of a car. Some assumptions are not taken into consideration like different friction of roads in rainy weather. One of the hurdles being using a vehicle whole ability and then infuse it with trajectories [19].

3.3 Risk Gauge

For inquiring and measuring the possibility of risk collision time, collision distance, reaction time are taken into consideration and some unusual situation. The problem in following this indicators is that it follows a constant speed, constant acceleration in straight road situation and might face challenges in roads with arc or banking [4, 19].

3.4 *Adjacent Vehicle Sense*

The sensors treat the vehicle as isolated object which limits it vehicle perception to its individual sensors. The vehicle either assume traffic conditions flawlessly or take help from the sensors to know the near ideal scenario of the obstacles. These approaches might face trouble as autonomous vehicle have a narrow vision with blind spots more when compared to that of the human driven vehicle [4, 19].

The open issues in self driving car includes its inability to take faster decision like humans based on the changing dynamic situation on road taking into consideration moral dilemmas. This gap is very much reduced with advent of technologies based on machine learning, bigdata analysis and Internet of Things and with faster processor. Self-driving car can now be trained as same or better than humans.

4 Test Results

The result of the lane change maneuver was performed on a plain area having 3.5 m road width and was led for various velocities at 10 and 15 m/s, separately. The outcomes show that the most extreme worth of sidelong speed increase for the vehicle is directed underneath the basis of the given parallel speed increase bound [18].

A Simulink-CarSim which is a tool for simulating vehicle's dynamic movements was used to test the LCM algorithm. Based on the LCM algorithm, the vehicle's simulation was regulated for various velocities and the results are very significant and are shown in the Fig. 7.

A Simulink-CarSim which is a tool for simulating vehicle dynamics movement was used to test the LCM algorithm. Based on the LCM algorithm, the vehicle's simulation was regulated for various velocities and the results are very significant and are shown in the Fig. 7.

Figure 7a shows the direction of the lane change [18]. Figure 7b shows the simulated lateral acceleration The Table 2 shows the number of iterations and the time taken in milliseconds [18].

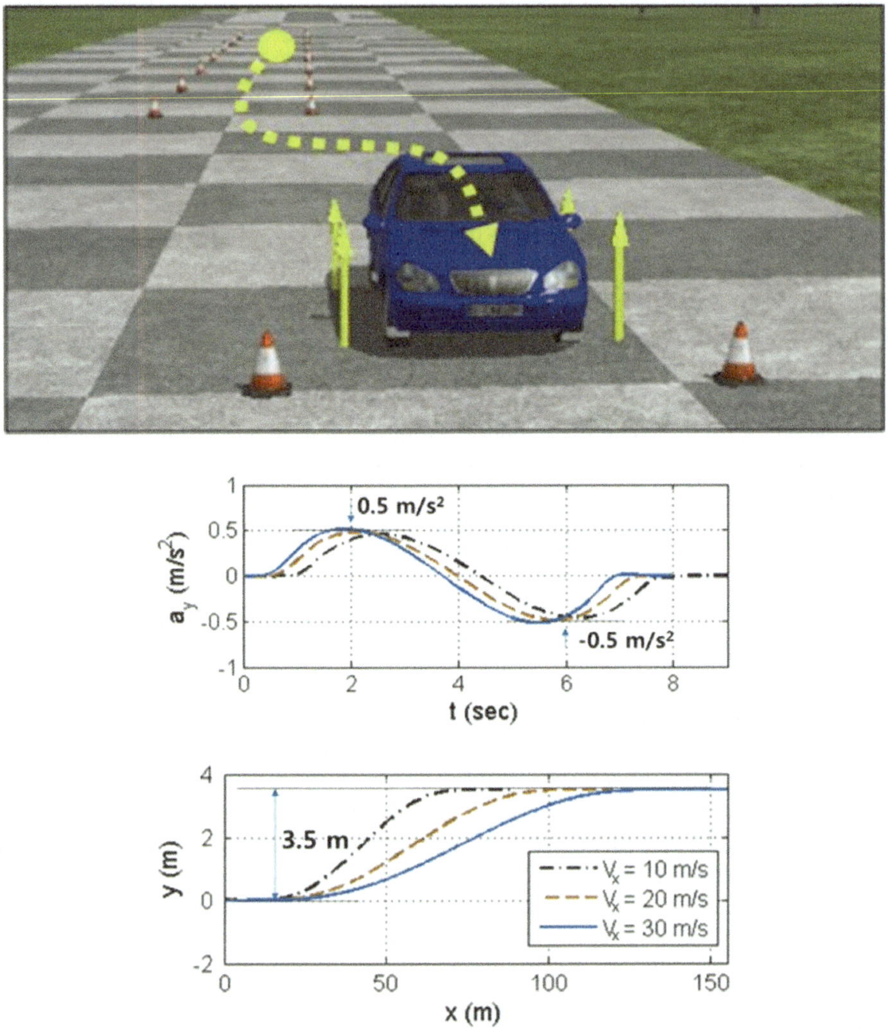

Fig. 7 A screenshot of Simulink-CarSim wile executing the LCM algorithm [18]. **a** Shows the direction of the lane change [18]. **b** Shows the simulated lateral acceleration [18]

Table 2 Shows the results of the number of iterations and the time taken in milliseconds for generating a LCM path, where V_x is the various velocities in m/s [18]

V_x (m/s)	Number of iterations	Time taken (ms)
10	13	218
20	26	267
30	39	322

4.1 *Grand Cooperative Driving Challenge*

The expansion towards maneuvering is wanted to oblige normal roadway maneuvering, like consolidating, and also to empower metropolitan applications. For this purpose, a layered control engineering is embraced. In this engineering, the strategic layer has the collaboration conventions, depicting the remote data trade to start the vehicle moves, upheld by a novel remote message set, though the functional layer includes the vehicle regulators to understand the ideal moves. This engineering was the reason for the Grand Cooperative Driving Challenge (GCDC), held in May 2016 in The Netherlands. The GCDC gave the chance to the participants to helpfully execute an expressway path reduction situation and a metropolitan crossway passing situation [1, 11].

4.2 *Second Autonomous Self-driving Vehicle Competition*

This event was organised on September 2012 at Namyang Research Center of Hyundai Motor Company, Korea and the results of two days fundamental rivalries were amassed as a proportion of 3:7 to decide a position. The contest course was a 3.4 km blended in with clear and unpaved streets while the opposition mission comprised of nine missions altogether like overwhelming lethargic vehicles, startling impediment evasion, detecting of traffic signals in the intersection, and stopping, detecting of traffic signals in the intersection, detecting of both ways turn, and detecting of pedestrian in line required progressed picture handling capacity while hindrances aversion and diversion required complex way age and following. What's more, sudden obstruction and overwhelming lethargic vehicles required capacities of detecting of progressively moving deterrents and producing an aversion way progressively. The last mission was to stop automatically at the assigned stopping region demonstrated by a number in the street sign. The unpredictable snag aversion mission was to pass the driving street that was obstructed with cones while the diversion mission was to reroute the current way. It was thus confirmed that path planning was effectively done [9]. Figure 8 shows the daigrammatic representation of self-driving vehicle competition organised by Hyundai Motor Company. Table 3 shows the present state of art based on self-driving cars.

5 Discussion

The development of self-driving cars is one of the most contemporary technologies in the world of Machine Learning (ML). These cars can find the best routes without crashing with the obstacles, recognizing the objects, interpreting its surroundings, making necessary decisions with the help of object detection and object classification

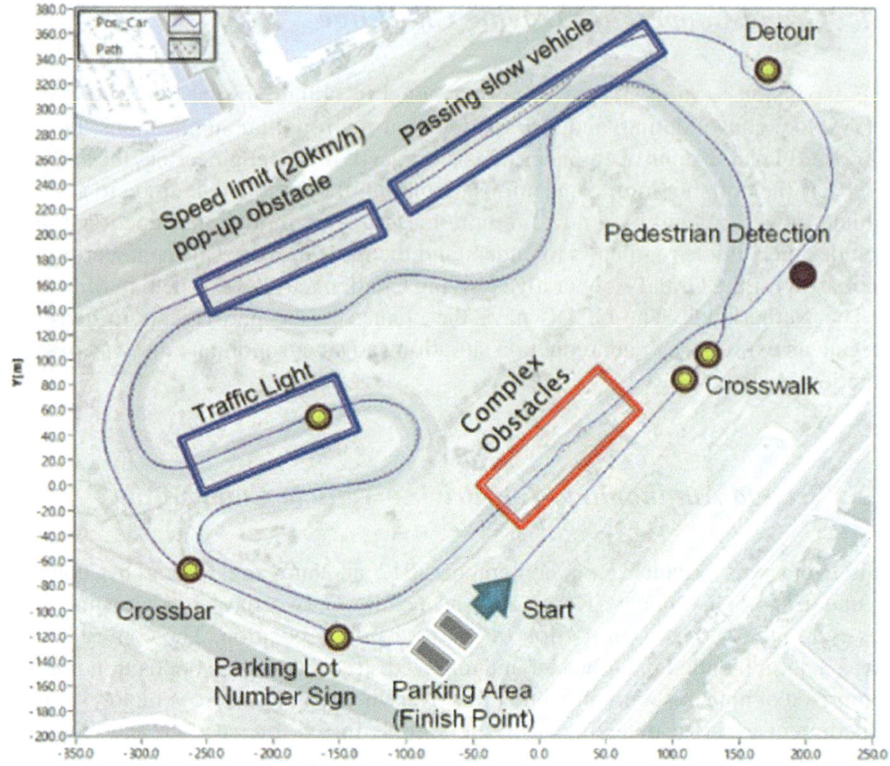

Fig. 8 Course of the self-driving vehicle competition organised by Hyundai Motor Company [9]

algorithms. Self-driving cars are equipped with high-end technologies. With ML, they can be taught the shapes of different objects. The three major sensors that are used in self-driving cars are cameras, radar and liDAR and these sensors allows the car to get a clear view of its environment and also in identifying the location, speed, and 3D shapes of the objects. These cars are also loaded with many safety features like the Advanced Driver Assistance Systems (ADAS) which distinguish nearby hindrances or driver's blunders, and react accordingly. The collision avoidance system warns the driver in case of any adverse circumstances and activates the emergency braking system to full potential and also manoeuvres the car in such a way that it avoids the collision. There are many mechanisms by which they do so, and putting them together allows these cars to judge and drive itself just like a normal human being.

Table 3 Summarization of study based on self-driving cars

Sr no	References	Methodology	Output	Application area
1	[1]	In the road merge technique, the vehicle either enters one ongoing traffic in the highway or exits out from the current traffic flow. It is done by changing its speed and maintaining a safe distance from the other vehicles. The merging vehicles are considered as an obstacle by the obstacle avoidance (OA) controller and therefore it manages to keep itself away from it by maintaining a safe distance	This methodology for maintaining a safe distance and arranging itself makes it even better against the speed distribution brought about by the other traffic, or by not occupying the merging strategy. This methodology is very helpful in cautiously changing the lanes in the highways	Maneuver planning in self-driving cars
2	[1]	In the intersection passing scenario the situation is made safe by attaining the same virtual platooning approach. Each vehicle figures out a virtual inter-vehicle distance between itself and the target vehicle. At the crossway, the vehicle sends the intersection's computer through wireless communication to tell it that a vehicle is coming. And afterwards the wireless communication again holds its entrance space, and time in the crossing point. The computer in a split second evaluates every vehicle's position, speed, and size comparative with different vehicles coming from all directions and simulates all possible paths of the vehicle	With a slot-based traffic design, vehicles can change pace steadily, to permit new vehicles to slide into their foreordained spots in the line. The vast majority of the methodologies for traffic sign identification are model-based and it uses basic features such as colours, shapes, and edges. This methodology was tested in the second autonomous self-driving vehicle competition held at the Namyang Research Center of Hyundai Motor Company, Korea and it was verified that path planning was successfully done	Path planning in self-driving cars

(continued)

Table 3 (continued)

Sr no	References	Methodology	Output	Application area
3	[9]	The crash detection algorithm is divided into two parts: the first part detects any obstruction present in the path in between the vehicle's position to the control point. Whereas the second part detects the presence of the obstruction by expanding the way to one side and left side to make an impact discovery limit dependent with the understanding that vehicle follows the way accurately	It helped in lessening the quantity of mishaps and fatalities on the present streets. One of the most encouraging components is forward-crash warning (FCW). It works by using the sensors or cameras to screen the distance between the vehicle and one in front. If there is any danger of hitting that vehicle, it alarms through an audible and visual admonition. It can likewise pre-charge the brakes to give most extreme slowing down capacity to the driver, and a few frameworks fix the safety belts in anticipation of an impact	Collision Avoidance in self-driving cars
4	[18]	The lane change maneuver works by deciding the wanted direction by accessing the suitable acceleration with the help of Bezier curve. After increasing the acceleration, the LCM algorithm changes the lane by confirming whether the surrounding paths are clear or not. Finally, the generation of the new path and changing its course into it is taken care by the trajectory planning system	The lane change maneuver algorithm comes up with a very pleasant and also safe and secure experience while changing the lanes because it easily skips the sudden increase in speed regardless of the different speed conditions	Maneuver planning in self-driving cars
5	[10]	In the Trajectory planning algorithm the planning is done by first of all finding the best geometric path and the best sequence of actions through incremental search for the vehicle. Next, it searches the best action from the multiple local search and also the best trajectory to follow through the optimisation of a geometric curve, according to given constraints allowing it to perform the best maneuver	This methodology is very beneficial in detecting the best directions in order to have a better path for the car. This feature is very advantageous because it is time and fuel efficient. Also, with the help of this methodology, the journey can be much smooth while travelling on bad roads or during off-roading	Motion planning in self-driving cars

(continued)

Table 3 (continued)

Sr no	References	Methodology	Output	Application area
6	[10]	Model Predictive Control (MPC) is an alternate methodology for trajectory planning, which joins parts of control engineering inside the planning module. Inside the MPC, the vehicle has a unique model which helps to input the inspected sample about the future advancement of the vehicle's movement from the controller. And the streamlining issue of tracking down the best direction for the vehicle is solved with the help of this dynamic model and the controller inputs	MPC was tested inside a driving passage on a linear bicycle model with direct tyre attributes which additionally considers horizontal and incline elements. The outcomes of this experiment presented a slight lateral acceleration which was more comforting to the passengers as compared to that of spline and lattice methods in which more bump was felt. The only drawback of this technique is that it gets even harder to optimise the trajectories when the variables used to model the vehicle is more. This methodology is more applicable on highways and its performance does not depend on the number of obstacles	Motion planning in self-driving cars
7	[10]	Akima splines curve is a more thorough methodology, as far as maneuvers and traffic rules are considered. In this methodology, the addition of Akima Splines curve helps in producing the trajectories which follows the checkpoints formed by the path planning module. This approach also produces stop signs, traffic signals, turns, path changes, road convergences, give-way signs, reverse driving and dead-ends	The trajectories were assessed by their distance and the time taken to get to their next designated spot, and the crashes were also estimated with the obstructions. Illustrations presented in a simulated surrounding as well as in real-world have confirmed that this technique is adequate for a safe and independent driving	Motion planning in self-driving cars

6 Conclusion

The tests have proved that self-driving cars are very practical and safe making them very efficient. These automated vehicles have numerous benefits namely traffic efficiency, better access and mode of transportation, prevention of crashes, societal cost-saving and are environment friendly.

References

1. Ploeg, J., Semsar-Kazerooni, E., Morales Medina, A.I., De Jongh, J.F.C.M., Van De Sluis, J., Voronov, A., Englund, C., Bril, R.J., Salunkhe, H., Arrue, A., Ruano, A., Garcia-Sol, L., Van Nunen, E., Van De Wouw, N.: Cooperative automated maneuvering at the 2016 grand cooperative driving challenge. IEEE Trans. Intell. Transp. Syst. **19**(4), 1213–1226 (2018). https://doi.org/10.1109/TITS.2017.2765669
2. Huang, Z., Chu, D., Wu, C., He, Y.: Path planning and cooperative control for automated vehicle platoon using hybrid automata. IEEE Trans. Intell. Transp. Syst. **20**(3), 959–974 (2019). https://doi.org/10.1109/TITS.2018.2841967
3. Chen, Z., Huang, X.: End-to-end learning for lane keeping of self-driving cars. In: IEEE Intelligent Vehicles Symposium, Proceedings, IV, pp. 1856–1860 (2017)
4. Kuwata, Y., Fiore, G.A., Teo, J., Frazzoli, E., How, J.P.: Motion planning for urban driving using RRT. In: 2008 IEEE/RSJ International Conference on Intelligent Robots and Systems, IROS, 1681–1686 (2008). https://doi.org/10.1109/IROS.2008.4651075
5. Badue, C., Guidolini, R., Carneiro, R.V., Azevedo, P., Cardoso, V.B., Forechi, A., Jesus, L., Berriel, R., Paixão, T.M., Mutz, F., de Paula Veronese, L., Oliveira-Santos, T., De Souza, A.F.: Self-driving cars: a survey. Expert Syst. Appl. **165**(July 2020), 113816 (2021). https://doi.org/10.1016/j.eswa.2020.113816.
6. Hubmann, C., Schulz, J., Becker, M., Althoff, D., Stiller, C.: Automated driving in uncertain environments: planning with interaction and uncertain maneuver prediction. IEEE Trans. Intell. Veh. **3**(1), 5–17 (2018). https://doi.org/10.1109/TIV.2017.2788208
7. Fassbender, D., Heinrich, B.C., Wuensche, H.J.: Motion planning for autonomous vehicles in highly constrained urban environments. In: IEEE International Conference on Intelligent Robots and Systems, 2016-Novem, pp. 4708–4713 (2016). https://doi.org/10.1109/IROS.2016.7759692
8. Frazzoli, E., Dahleh, M.A., Feron, E.: Real-time motion planning for agile autonomous vehicles. In: AIAA Guidance, Navigation, and Control Conference and Exhibit, vol. 25, no. 1 (2000). https://doi.org/10.2514/6.2000-4056
9. Lee, U., Yoon, S., Shim, H., Vasseur, P., Demonceaux, C.: Local path planning in a complex environment for self-driving car. In: 4th Annual IEEE International Conference on Cyber Technology in Automation, Control and Intelligent Systems. IEEE-CYBER 2014, pp. 445–450 (2014)
10. Katrakazas, C., Quddus, M., Chen, W.H., Deka, L.: Real-time motion planning methods for autonomous on-road driving: State-of-the-art and future research directions. Transp. Res. Part C Emerg. Technol. **60**, 416–442 (2015). https://doi.org/10.1016/j.trc.2015.09.011
11. González, D., Pérez, J., Milanés, V., Nashashibi, F.: A review of motion planning techniques for automated vehicles. IEEE Trans. Intell. Transp. Syst. **17**(4), 1135–1145 (2016). https://doi.org/10.1109/TITS.2015.2498841
12. Cheng, D., Qi, H., Zhao, Y.: Analysis and control of general logical networks—an algebraic approach. Annu. Rev. Control. **36**(1), 11–25 (2012). https://doi.org/10.1016/j.arcontrol.2012.03.002

13. Broggi, A., Medici, P., Zani, P., Coati, A., Panciroli, M.: Autonomous vehicles control in the VisLab intercontinental autonomous challenge. Annu. Rev. Control. **36**(1), 161–171 (2012). https://doi.org/10.1016/j.arcontrol.2012.03.012
14. Li, J.Q., Kong, N., Hu, X., Liu, L.: Large-scale transit itinerary planning under uncertainty. Transp. Res. Part C Emerg. Technol. **60**, 397–415 (2015). https://doi.org/10.1016/j.trc.2015.09.010
15. Villagra, J., Milanés, V., Pérez, J., Godoy, J.: Smooth path and speed planning for an automated public transport vehicle. Robot. Auton. Syst. **60**(2), 252–265 (2012). https://doi.org/10.1016/j.robot.2011.11.001
16. Luo, Y., Xiang, Y., Cao, K., Li, K.: A dynamic automated lane change maneuver based on vehicle-to-vehicle communication. Transp. Res. Part C Emerg. Technol. **62**, 87–102 (2016). https://doi.org/10.1016/j.trc.2015.11.011
17. Claussmann, L., Revilloud, M., Gruyer, D., Glaser, S.: A review of motion planning for highway autonomous driving. IEEE Trans. Intell. Transp. Syst. **21**(5), 1826–1848 (2020). https://doi.org/10.1109/TITS.2019.2913998
18. Bae, I., Kim, J.H., Moon, J., Kim, S.: Lane change maneuver based on Bezier curve providing comfort experience for autonomous vehicle users. In: 2019 IEEE Intelligent Transportation Systems Conference, ITSC 2019, pp. 2272–2277 (2019)
19. Jin, C.J., Wang, W., Jiang, R., Zhang, H.M., Wang, H., Hu, M.B.: Understanding the structure of hyper-congested traffic from empirical and experimental evidences. Transp. Res. Part C Emerg. Technol. **60**, 324–338 (2015). https://doi.org/10.1016/j.trc.2015.09.007

Autonomous Vehicle Assisted by Heads up Display (HUD) with Augmented Reality Based on Machine Learning Techniques

S. Murugan, A. Sampathkumar, S. Kanaga Suba Raja, S. Ramesh, R. Manikandan, and Deepak Gupta

Abstract The safety in driving is improved and driving workload is minimized, the provided information is understandably and the cognitive load on the driver is low. For the autonomous vehicle, machine learning-based AR-HUD (augmented reality based Head-up display) is used. In this paper the machine learning-based AR-HUD has been used for autonomous vehicles. The process of object detection and collected HUD data classification has been done by the proposed model. Determining the present state of vehicle has been validated based on the AR environment. The process of test and validation is an integral portion of a development cycle. Machine learning and deep neural network are used in this paper for lab & real-world T&V for ARHUD and autonomous vehicles. The results of simulation obtain the data gathered from the implementation of human and machine interface (HMI) to detect the object and to classify the objects in motion. Accuracy, precision, recall and F-1 score are the analyzed parameters for machine learning-based ARHUD. The simulation results obtained are accuracy of 98%, precision 94%, recall 92.3% and F-1 score 86% in comparison with CNN, ANN and SVM.

S. Murugan
Department of CSE, Sri Aravindar Engineering College, Villupuram, Tamil Nadu, India

A. Sampathkumar
Department of CSE, Dambi Dollo University, Dembi Dolo, Ethiopia
e-mail: dr.sampathkumar@dadu.edu.et

S. Kanaga Suba Raja
Department of CSE, Easwari Engineering College, Chennai, Tamil Nadu, India

S. Ramesh
Department of ECE, Vinay Foundation for Science, Technology Research, Guntur, Andhra Pradesh, India

R. Manikandan (✉)
School of Computing, SASTRA Deemed University, Thanjavur, India

D. Gupta
Maharaja Agrasen Institute of Technology, Delhi, India
e-mail: deepakgupta@mait.ac.in

A. E. Hassanien et al. (eds.), *Virtual and Augmented Reality for Automobile Industry: Innovation Vision and Applications*, Studies in Systems, Decision and Control 412, https://doi.org/10.1007/978-3-030-94102-4_3

Keywords Autonomous vehicle · AR-HUD · T&V · HMI · Object detection · Classification

1 Introduction

Autonomous driving systems are highly complicated and needed to be efficiently tested previous to deployment. The safety of autonomous driving system's assurance in serious circumstances is difficult task, and the process and concepts of testing are deliberated. At present, numerous works on AD are performed by ADAS to automated and autonomous driving tests. Though, the methodologies for testing the autonomous vehicle completely is till required highly in the entire process of development, incorporating development of functions and testing, integration of the system and its verification, test drive and authentication etc. Quick expansion of AV has worn excessive worldwide attention in current years. The modern transportation systems with promising AVs are predicted for addressing several difficulties in long-standing transportation concerning congestion, safety, parking, energy conservation, so on. Possibly, several technologies of AV in lab, closed-track and public road tests produced improvements that are noticeable in bringing AVs in the applications of real world. Advanced Driver Assistance Systems (ADAS) was increasingly linked to semi-automated driving concept with prevailing wisdom assisting drivers in the process of driving. Lane departure warning and blind spot alerting are some of the applications linked to these systems. The safety enhancements of vehicle systems are adapted, improved and automated with the error reduction related to human drivers. Many methods of AI produce promising results for recognition of environment by AVs and vehicle propelling for suitable decisions. AI applications in AV are reviewed for the development of AV like planning motion, perception, making decisions and validation of safety. Nevertheless, the existing system provides an inclusive review from the applications of AVs related by AlphaGo in which the humans are defeated.

The severity of accidents is reduced by various kinds of automation of vehicles and mobility increases with two core requirements of increasing global population. Though this is not vibrant complexity grasp elaborated with different automation types and development of wide array depending on the above systems which support the needs of mobility. For the efficient development and deployment of the system, the safety of the drivers and mobility is enhanced by highest automation degrees, understanding of the consumers, trust and system desirable requirements are used for supporting the marketplace.

Nowadays, autonomous vehicles (AVs) are established rapidly and gained high consideration worldwide. The innovation of modern system of transportation was promised by AVs and anticipates the challenges in long-standing transportation associated with safety, congestion, parking, conservation of energy etc. Public road tests closed track tests are the lab test for the technologies of AV and produced relevant improvement takes AV into real-world applications. This enhancement in AV requires considerable investments and promotions by various stakeholders like agencies of transportation, information technology (google, baidu etc.) (IT), network companies in transportation (e.g., Uber, DiDi, etc.), producers of automobiles (e.g., Tesla, General Motors, Volvo, etc.), manufacturers of semiconductor/ chip (e.g., Intel, Nvidia, Qualcomm, etc.), etc. However, the self-driving or AV's concept is familiar. Pioneering data collection and processing technologies and artificial intelligence are combined to process the AV research to extraordinary heights. The usual knowledge of previous (semi-)automated concept of driving was greatly associated with advanced driver assistance systems (ADAS) assisting the processing drivers while driving. Alteration of blind spots and Lane departure warnings are the application related to the system. Technologies of augmented reality enhancements are grown rapidly nowadays concerning hardware and software, specifically for mobile devices and head-mounted displays. The comfort of drivers and safety is improved by techniques of AR in automotive sectors and next-generation head-up display's (HUDs) form are applied by infotainment platforms are also acted by this technique. HUDs are equipped typically by today cars with having restricted horizontal field of view (FoV) of up to 5°. Implementation of virtual image on car windshield with conventional 2D projection having virtual image distance of up to 3 m in these systems. Enlargement of virtual image's size, characteristic representation of information like HUDs includes the instrument cluster's replication having the extension of restricted navigation. Automotive sector's recent enhancements are HUD's development having the support of AR, affording the capability for merging the related driving having actual ("real") scene. Several sensors produce this information to help the driver in collision avoidance or the reception of extra information for assisting drivers. As per the immersion level given to the users the categorization of VR is performed. Non-immersive visualizations like the setup of desktop monitors are simple and economic for implementation. Accomplishment of Semi-immersive VR having head-up displays (HUDs) produces the real-world augmentation having digital overlay. Cave automatic virtual environments (CAVEs) and head-mounted displays (HMDs) are contrasted comprising projection are casted on walls providing immersion's high level to the user because of real environment's movement in the background. The vehicle automation's improvement from manual and driving assistance towards fully and highly automated vehicles with the requirement having classification requirement are sharing roads and this is usual. Thus, a usual language was introduced by SAE On-Road Automated Vehicle Standards Committee for the automation level identification. Vehicle automation's every level produces both challenges and opportunities for practitioners and researchers. The awareness situation and concerns of safety are needed for addressing the level of higher automation for lower levels and new interaction was brought and experiencing passenger to car. Subsequently, the features of automated driving research are carried out by VR technology and experiences a safer and controlled environment. Moreover, highly automated vehicle's present restricted availability makes the researchers investigate AD scenarios through other means like the simulation of virtual reality. Moreover, the advancements of technology and VR equipment's affordable increment facilitates the uses. The subjective participants showed studies and responses of psychology in the environment of virtual reality are tightly coupled for experiencing and real world set up behavior. This system is motivated towards the adaptation and improvement in few safeties enhanced vehicle systems in which the errors are reduced linked with human drivers [1] as the drivers are still needed to carry out the task of driving, and automation at lower level are considered in these systems as per the Automotive Engineer's classification (SAE) Society [2, 3]. The human driver's involvement is reduced and this system is to start to proceed from conditional automation level3, high automation level 4 and at level 5 there is full automation finally. In fifth level, a vehicle is predicted to drive on its own suitable for all circumstances of environment. The "perceiving", "thinking", and "reasoning" are the capabilities of human driver which is needed by the purely self-driving level. In recent years the enhancement in AI tends to achieve consistency in-between AI and AVs to meet the requirements [4]. Instead of performing a critical role in system acceptance depending on autonomous agents, incorporating autonomous vehicles (AVs), it is very difficult to develop trust among humans and machines. As per the survey in 2017 done by Pew Research Center, people are interested to go on the vehicle which is without drivers. From the literature it is noted that partial autonomous driving case produces undesirable feedback by the Human–Machine Interaction (HMI) design adoption (HMI) which produces the feedback of acting car. Through the establishment of cooperative relations among the passengers/drivers and vehicles, HMI will affect positively the acceptance and also AD solution's technological advancements. Inappropriately, the combined approach's application for designing UX and evaluating AD systems is difficult. For example, quantitative assessment is focused on specific design of user interface techniques for measuring the performance of driver in particular driving job that is difficult to reuse when the automation is at specific level with passengers and no drivers. Likewise, the questions obtained from the previous experience are highly suitable when the feedback collection is concentrated with great aspect amount which is denoted as the trust at perceived level. The behavior of the vehicles is intelligent and is developed by driving simulators and is not applied directly to UX study, where the shifting of focus happens, for example, interior part of vehicle having the interaction with it instead of external fidelity factor which affects the decision. A methodology of HMI and IVs are presented in this paper for evaluation of trust level among the passengers with the consideration of designing the AD system with probable interface. The methodology planned depends upon the immersive Virtual Reality (VR) platform, which is established through the existing driving simulator grounding. Though several scenarios are applied in this technology by focusing L4 and L5 levels of automation for the experience of the passengers are characterized by representing the configurations in the aspects of trust and comfort and are highly difficult. A virtual AD system was created that permits the user to feel the simulated ride in an urban virtual environment, to face numerous dissimilar situations. Specifically, AI's success in several applications like AlphaGo possesses improved research in AI leverages in the development of AV. Particularly, Various problems in AVs are tackled by several studies in Deep learning (DL) e.g.; accurate recognition and obstacle location on roads, suitable in making decisions (e.g., steering wheel control, deceleration/ acceleration), etc. On the whole, several approaches of AI provide a promising AVs solution in environment recognition and vehicles are propelled with suitable decisions. Some studies are reviewed particularly in the AI application in a particular component related to the development of AV, e.g. perception, planning motion, making decisions and validating safety.

Motivation

The main objective is to evaluate proposed method's appropriateness. Head-Up Display (HUD)-based interface for automated vehicles are designed which produces visual signals about the sensory vehicle and system planning. It is shown that the information is provided with the car's action that is crucial for eliciting AV's trust, but very small evidence of experiments are available for the determination of some information is provided to the passengers in best way [5, 6]. This approach is applied for the aforementioned scenario, and quantitative and qualitative proof was obtained with complete picture surrounding the vehicle. Furthermore, exposure of interface providing the content of higher information users incorporated in this study which is highly interested in checking real system AD. Future HMI designs are driven by providing interesting insights, this approach's efficiency is confirmed by the results obtained representing UX's probable facets that were evaluated by the experiments of techniques. For UX's assessment both cognitive and affective factors are considered, by feedback integration depending on the queries of previous experience, gathered information from signals of physiology. The measurements of stress levels are focused in this paper specifically to evaluate the degree of connection and safety with the vehicle. Nevertheless, future extensions of detecting the state of emotion are supported by the design of proposed methodology. The latter state is more realistically measured by VR as compared to other conventional scenarios of simulation.

Contribution

The contribution of this paper is to develop autonomous vehicle assisted with head-up display with analyzing the current state of the vehicle. By collecting these data the object detection and classification of moving scenes has been carried out using RCNN and AlexNet CNN.

Related works are presented in Sect. 2. Section 3 shows proposed model for autonomous vehicles assisted by AR with HUD model and data classification. Evaluation criteria are discussed in Sect. 4. At the end, the conclusion is presented in Sect. 5.

2 Related Works

Nowadays, Augmented reality is established by several automotive manufacturers because of its accessibility and generation of innovative solution potential. Human Machine Interaction (HMI) enables AR where HMI's virtual components are introduced and overlaid on the real world, by illusion created for reality enriched. Some of the information is obtained by allowing the users as it requires few procedures or aspects straightly in the environment of working with the possibility of enriching sensory perception of person. Depending on the definition of Milgram of reality-virtuality continuum, mixed reality's (MR) part is represented by AR that merges the virtual and real world. The AR range's applications from the activities of industry to everyday life, in different domains such as entertainment, robotics, manufacturing, education, healthcare, military, etc. Industrial AR with impact and implications are addressed in many extraordinary studies and also direction development and difficulties in this AR field arise the emerging possibilities in each automotive industrial area because of its potential as an intuitive and interactive interface. The AR usage in the automotive industry was grown in several areas [7]. For example, the experience of the user with HUDs is improved substantially by AR. From military aviation, the display systems of HUDs are adapted for transmission of visual information from the visual field of the driver. All these are designed for presenting several necessary information in real-time for increasing safety. HUDs provide two ways: providing excessive information to the environment or few elements present already are highlighted. Video stream is overlaid by some of the interesting solutions over the car forward-facing of the driver. Conservation, restoring, diagnostics, examination or training are few other aspects contributed by AR. Before ordering car the customers are customized interactively by using AR as a marketing tool. The projector-based spatial AR is the basis for spot welding inspection, in which the few vehicle parts are tracked or procedures for maintenance are performed by head-mounted displays (HMDs). The workers and operators in the field of producing cars are proved to be efficient in human-centered technology. Various principles of design perspectives are implemented for AR HUDs in several studies. Therefore, FoV and brightness of the image are increased by utilizing micro-electro mechanical-system depending on digital light processing, or 2D computer-generated holography (CGH) and laser scanning. When holographic optical elements, waveguides applied or mirror-based projection's meta surface technologies instead to make AR HUD highly compact. Though AR content matching's enhancements are not addressed by this solution and this enhancement encloses visual conflicts avoidance in real scenes. The multi-depth images implementation for merging a virtual image's set having real scene is an efficient method for addressing this problem. At present, image plane's numbers are restricted and thus inadequate for natural overlapping. Moreover, for adaptation of mechanical steering, flickering will happen. The varifocal optical elements are used to rectify these limitations; though it is a challenging task for practical application because of the consideration like the needed dimensions/ size, curved windshields having compatibility, and large size eyebox. Further, this is an eventual resolution for addressing natural multi-depth image perception cues. Though, application of CGH is multifaceted concerning hardware and version and time of computation is needed additionally; and therefore it is not suitable for high-speed automotive solutions. An approach that bonds the gap between holographic 3D displays and conventional systems is the autostereoscopic 3D display's integration into the HUD. In spirit, binocular disparity is the basis for works: the depth's illusion was created by a pair of 2D images with slight offset features from each other, and the merging of brain with these images into perspectives of 3D within the depth range which is acceptable. The images of valid stereo-pair are observed by many visual positions which are formed by Autostereoscopic 3D displays. The perceived 3D image's resolution is decreased based on the viewing point number and this is the major limitation. Though, utilization of pixel resources is improved by a method for a huge viewing point's number was implemented by a light-field model. As eye-tracking is joined with direct light-field translated for valid stereo-pairs is impossible with a 3D image resolution significant loss. Though at last, binocular disparity is exploited, this display type was measured as a display of light-field as the stereo-pair images are allocated to uniform light rays distribution as per the model of light-field. Design of HMI was included in various driving aspects investigation of driving in the scenarios of simulation, dependent on performance and behavior of drivers as proxy for emotional and cognitive status. Standardized questions and also indicators like speed of driving, keeping lane, patterns braking etc. are included in experiments performed and relative or absolute validations are established in [8]. In AV's, driving part of human are reduced progressively making the assessment of behavior less applicable. In engineering, the cognitive and affective states of users are measured by using physiological signals [9]. Driving simulators of conventional methods and immersive driving simulators are investigated in [10]. Several studies support the conventional driving simulators with physiological signals relative validity, less abundant is available for the performance of driving [11]. Risk perception was highly correlated with the GSR alteration. Simulated driving and on-road conditions developed the relative validity are compared with mean HR and mean consumption of oxygen [12]. In VR environment which is immersive, physiological response is robust than the conventional driving simulator during the pilot study [13]. In real-life driving conditions, physiological signals are recorded to analyze the performance of driver characterization from that the stress level is measured for the detection of drowsiness in drivers and this was studied by Healey and his colleagues in [14]. A route is set is followed by the driver by recording ECG, EMG and GSR and the sessions of driving are recorded in video tape and stress-induced actions are observable for the visual inspection, these actions include turning head represented as a reference standard. Various stress levels with increased accuracy are distinguished by the authors and allowed to collect signals; Drivers are mostly correlated closely to the metrics of HR and GSR. These studies are conducted from the active driver's view and results are analyzed the effective and equal observation of passengers. Particularly, the neural network (NN) applications are investigated in [15] and AI approaches like machine learning are not focused and so it not available for public. Safety problems are increased by ML techniques uncertainty like training data incompleteness, shift in distribution, variations in the operational and training environment, prediction uncertainty. Lane detection and traffic light detection are the topics related to CNN model structures to achieve similar level accuracy. For example, rules of fuzzy logic and knowledge of road are combined for AV's navigation vision [16]. Then, adaptive randomized Hough transform algorithm (ARHT) was proposed in [17] for lane detection and its validity is proved as it is contrasted with genetic algorithm depending on the detection of lane for AVs. Detection and tracking of moving vehicles with autonomous driving robot Junior of Stanford are performed by Bayesian network and Rao-Blackwellized particle filter combination [18]. CNN method, for detecting 9000 objects with 40–70 frames per second (fps) in real-time was proposed by the researchers in [19] having the accuracy close to 80%, making it able for the detection of every essential thing for automation tasks in an onboard camera or video. Saliency analysis and edge detection are some of the techniques used for high-level feature extraction for identifying objects. Support vector machines (SVM) is one of the classification approach introduced for CNN-learnt features further classification. Extreme machine learning is introduced for the classification of deep perceptual features by CNN and performance of competitive recognition up to 99.54% was achieved on the German traffic sign's benchmark data [20]. Contrasting the perception of single object, sharing of knowledge is imposed by multi-task object perception for getting the solution of numerous correlated tasks instantaneously and it is used for boosting the performance [21]. The region-of-interest voting was introduced for the implementation of detecting multi-task objects depending on CNN and this approach is validated on vehicle datasets of KITTI and PASCAL2007. The detection of objects is optimized simultaneously by the multi-task combination formed based on Cartesian product was introduced in [22]. In this method, prediction of object distance for full considers the task dependency. Additionally, rather than the object detection like vehicles and traffic signs roads that are segmented semantically having the surface drivable act as a major role in the perception of AV. In an image every pixel to label class is linked by semantic segmentation. Environment with the context having better understanding is given to AVs. The architecture for special CNN encoder-decoder is utilized in [23]. Later the network processed the input image, computing the pixel-wise classification for the determination of every pixel label and 88% of prediction accuracy was reported for cars and roads is 96%. In case of direct perception, the mapping-related computation's complete section of AVs, the local map is not created locally or any plan of detailed trajectory is not performed. Therefore, the stage of mapping and localization's majority are skipped by direct perception and the speed of vehicle and angle of steering output are controlled directly. The framework of edge-assisted privacy-preserving raw data-sharing framework was presented in [24]. Initially, the additive secret sharing technique is leveraged for the encryption of raw data into 2 cipher texts and secure functions with two classes are constructed. The privacy-preserving convolutional neural network (P-CNN) was implemented by using functions. At last, features are extracted by P-CNN by cooperative execution of deploying two edge servers from 2 cipher texts for obtaining the similar detection of object outcomes as the new CNN. The model of VGG16 is adopted as a case study for illustrating the construction of P-CNN and the KITTI dataset is employed for the verification of this solution. From the results of experiments, similar classification outcomes are exactly offered by P-CNN having negligible error in VGG16 model and the overhead communication and the cost of computation on edge servers are lower than the solution that existed with no leakage of private information. For autonomous vehicles, a real-time pedestrian detection system using CNN was presented in [25]. Scratch is used for system design with no usage of any available normal module. This system's evaluation is performed by INRIA dataset, PETA–CUHK dataset and real time video input and the probable accuracy of recognition are achieved by tuning CNN parameters. 96.73–100% is the accuracy range obtained depending on the employment of dataset. A typical single-task network was extended by this innovative approach in which two tasks abilities are carried out in [26] with no many decoders and only a single-stream two-task network is used. Prediction of monocular depth and semantic segmentation are two output tasks for AD's visual perception. The semantic segmentation having the function of regression loss is solved by this method. The evaluation of experiments with single-task baseline and multitask network was performed. Uncertainty-aware end-to-end trajectory generation method depending on imitation learning was developed in [27]. The spatiotemporal features are extracted by the images of front-view camera for understanding scene, collision-free trajectories are generated in numerous seconds in future. From the results of experiments, it is suggested that under different conditions of lighting and weather, trajectories are generated reliably by this network in various urban environments like intersection turning and collision avoidance slowing down. Moreover, the tests in closed-loop driving are suggested as it produces better results of driving in cross-scene/platform as compared to the existing end-to-end control system. Errors due to off-center and off-orientation are recovered by this model and 80% of hazardous cases with increased estimation of uncertainty are captured.

Despite the efforts in existing methods, which addressed the questions of previous research. An organized review was conducted by the identification and key component discussion of problems and difficulties in the AI and AD approaches and also further enhancements are discussed.

3 System Model

The safety potentials are improved by In-vehicle or roadside AR-HUD systems to provide the drivers with advanced alterations in warning in the geometry of routes. Mark-on-windshield warning system for vehicles is proposed in this study and enables the drivers to view the objects that are possibly dangerous appearing in front of the vehicle by the direct info projection in the field view of drivers, with no eyes taken from the road. The safety information is provided conveniently by the drivers and the obstacles forwarded are avoided safely by the proposed system. The overall architecture for AR-HUD with object detection is given in Fig. 1.

The cars without drivers, aircrafts without pilots and various automatic systems, the performance of tracking are improved by establishing communication with other vehicles in AR systems and the observed prediction and critical event detection is better. Partnership establishment with some car in surroundings in driving is allowed by the vehicle is focused in the proposed system and it also provides the danger warning and manages the car in few situations. In driving vision is significant and worsens the safety of the driver if the visual attention is declined. Presentations of AR are investigated in this section and hazards which are concealed visually are presented in best way. Safe distance besides other vehicles are maintained for avoiding collisions, Different mounted devices are used in the detection of obstacles and provide information and warning about it to the driver in this proposed system.

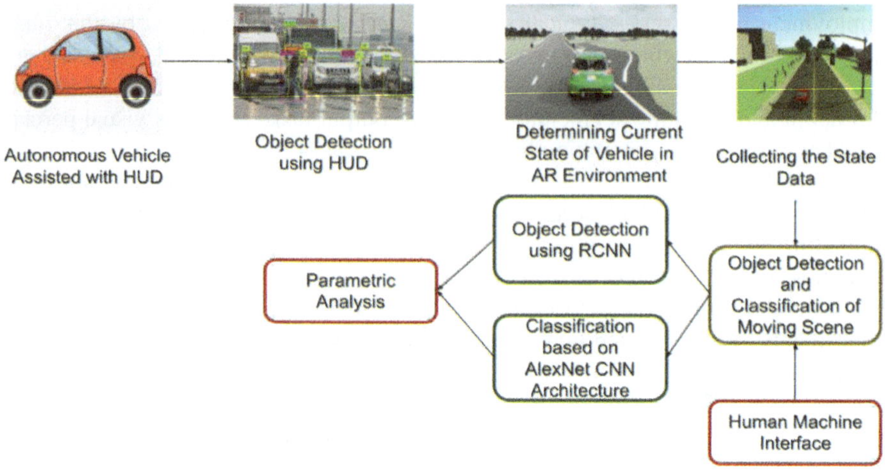

Fig. 1 Overall architecture for AR-HUD with object detection

On-board camera-based driver alert system was deployed in contradiction to signs of traffic like limit of speed, stop, unique danger signs etc. The lock-to-lock steering wheel travel was reduced by using this technique which is the vehicle speed function e.g.; if a car is stopped suddenly on the blind corner of V2V or around, and other cars are warned reduce speed previous the danger view. Due to these severe injuries are low and it is facilitated on some roads. The level of stress of every participant is decreased by the intelligent vehicle's support and leads to enjoyable overall experience of driving.

Technology and Setup: The VR system is implemented using the HTC Vive ecosystem (https://www.vive.com/eu/product/), by HTC Corporation, Taiwan. The resolution of 1080 × 1200 pixel/eye is features of Vive VR Headset covering horizontal 110° FOV at 90 Hz. Vive Base stations (based on the construction of Valve's Lighthouse method) emit the IR lasers which are leveraged by native position tracking. IR laser light and built-in sensors of headset are combined enabling six DOF outside-in chasing the head of user. Immersion of the foster is motivated towards the motion stimuli simulation in which the real vehicle experience is faced by the passenger or driver by using the platform of inertial motion. Atomic A3 Racing is used as a platform in this method, which is designed by Atomic Motion Systems, that gives support to yaw and pitch DOFs for simulation of motion. The acceleration perceived by the user is simulated by the implementation of strategy called tilt coordination motion simulation [7]. Shortly, the gravity acceleration vector's decomposition is used for perceiving acceleration which is given by the platform's coherent rotation. The VR coordinate system (headset is present in the center, i.e., viewpoint of the user) is applied with motion compensation, depending on rotation of current platform. For this purpose, seat is mounted with Vive Tracker along with headset. At last, in this virtual environment, visualization of user's hands improves the presence of sense, hands of the user provide virtual replication incorporating the creation of articulated fingers and tracking it by utilization of Leap Motion Controller device with headset attachment. The open source Simulator Vehicle project by the GENIVI Alliance [29] forms the basis for the implementation of vehicle simulator. The reasons for selecting GENIVI among many probable alternatives are: HMI design was supported; new features are allowed to be added; simulation of intelligent traffic modules are provided already; Functionality of basic auto-driving was included for the vehicle of the user; some scenes for driving are provided with its individual rigid body physics-based controller. The major actions performed for adapting GENIVI to this work involve: available features porting to VR; platform of motion are integrated; custom AD controller implementation. Essential activity of latter was considered in the basic study, for judging built-in controller is not enough to deal complicated events which are unpredictable (sudden pedestrian crossing so on). (1) VR porting: Implementation of VR with support was simplified by GENIVI depending on Unity game engine, in the applications of VR creation are allowed natively for HTC Vive. User with virtual accommodation in any seat present in virtual vehicle was allowed in this implementation. Land Rover L405 and a Jaguar XJ are Built-in vehicles are intended for the non-immersive simulation design. VR-based interaction is used in the construction of new vehicles with the focus on interior of vehicle's visual fidelity. At last, adding the user support (2) Integration of motion platform: developing extra software module for the integration of motion platform. The values of acceleration are the inputs received by the module which is computed by the tracking point of seat by the engine of simulation in physics and results in the driver of proprietary platform (AMS Symphinity), in which the pitch angles and coherent tilt are remapped and the platform is applied with this consequently. In the same manner, other platforms of motion are integrated. (3) AV simulation: simulated VR-based scenarios are considered in this method, escorted by appropriate tools for measurement tools, instead of contributing to the improvement in the existing control sub-systems of AV. In GENIVI, the basic available AD functionality was prolonged for handling the interesting situation. Attention towards reproducibility, in the process of simplicity preservation. Implementations and leveraging are highly sophisticated e.g. virtual sensors of vehicles providing data are integrated in future.

Visual cues are ensured by AR-HUD and it is believed as highly important and the conveyed information is consistent through the commercial products of DA because the users are frequently acquainted through it. Though it was observed as critical for providing the information which defines the sensory capabilities of vehicles and thus awareness situation of the user was improved. At last, AD systems L4 and L5 are focused in this work, planning functions of the vehicle information were required to be also delivered.

Detection of Traffic sign: This step considers the traffic signs of distinctive features. Particular shapes and colors are used in the normalization of traffic signs, and the feature usage is convenient to take the candidate sign decision. The motivation of this detection stage is identifying the region of image that possesses traffic signs. The methods of fast detection are used for ensuring the performance of high system. This method is the outperforms detection of objects for the applications in real-time i.e., the framework of Viola-Jones face detector. The traffic signs are detected by system's 1st phase using scanning window with Haar cascade detector for the input stream's every image (target images) and eliminate the images without objects. However, Pixel noise and low contrast are avoided in the detection of traffic signs by the texture-based detector, in which ROI is post filtered to provide the conclusive detections.

More sophisticated and complex algorithms are significantly required by the poses in camera to find the problem in marker-less AR-HUD, e.g. mapping disparity, detecting feature, extraction, classification and matching. The calibration of Camera permits virtual world's combination and objects of real world in an individual display. The parameters of camera and the camera's relative information of position corresponding to the element, calculates the element's relative position on a screen for videos or images.

The transformation of projection is the mathematical model utilized represented in Eq. 1 where homogeneous scale factors unknown a priori is denoted by λ, 3 × 4 projection matrix is given by P, image feature's homogeneous coordinates is x = (x, y), In world coordinates, feature point's homogeneous coordinates is given by X = (X, Y, Z), the intrinsic parameters of camera in the matrix is denoted by K ∈ R3 × 3 as well as camera matrix, the [R|t] is the extrinsic parameter's joint rotation-translation matrix, 3 × 3 rotation matrix is R = [rxryrz] and camera translation is T = [t] was expressed in Eq. (1)

$$x = PX = K[R \mid t]X \tag{1}$$

The matrix of projection is provided in Eq. (2), key is created by the realistic augmented scene is P utilizing the camera's intrinsic parameters, the video frame's dimensions, and the close and distant clipping plane's distances from the projection center. Previously the knowledge of intrinsic parameters is derived in this method and these parameters are not altered, and this is sensible in many cases.

$$P = \overbrace{K}^{\text{Intrinsic Matrix}} * \overbrace{[R \mid t]}^{\text{Extrinsic Matrix}}$$
$$= \underbrace{\overbrace{\underbrace{\begin{pmatrix} 1 & 0 & x_0 \\ 0 & 1 & y_0 \\ 0 & 0 & 1 \end{pmatrix}}_{\text{2D Translation}} * \underbrace{\begin{pmatrix} f_x & 0 & 0 \\ 0 & f_y & 0 \\ 0 & 0 & 1 \end{pmatrix}}_{\text{2D Scaling}} * \underbrace{\begin{pmatrix} 1 & & s/f0 \\ 0 & 1 & 0 \\ 0 & 0 & 1 \end{pmatrix}}_{\text{2D Shear}}}} * \underbrace{\overbrace{\underbrace{(I \mid t)}_{\text{DD Translation}} * \underbrace{\left(\frac{R \mid 0}{0} \mid 1\right)}_{\text{3D Rotation}}}^{\text{Extrinsic Matrix}}} \tag{2}$$

Virtual objects are flawlessly integrated into real-world, desired environment is recognized and tracked by AR system the with its capabilities. Though, challenges are still present in the marker-less techniques and possibly explored, for developing a unified AR knowledge. Within the view of final augmentation, orientation, position, and scale with no visual inconsistency are present in the real objects are used in the alignment of the virtual object. The host path predicted and the traffic signs expected positions in the surrounding of the vehicle is in three-dimension model, then image plane projected results in ROI. For the drivers, the safety precautions are served by this AR technology having the impairments of visual attention.

Once the projection is transformed, the current state of the vehicle in AR environment has been collected. In each step of time, the vehicle's current state computes the target path and is described below. The torques are produced for vehicle guiding, in which the vehicle is apart from the safe state boundaries, to take the vehicle back to a safe state at a provided position target; it is needed to calculate the target path. The point, position target is the narrow road or the tight curvature inside the area sensed. The target path's smoothness is significant due to smoothness lacking in force causes a discomfort sensing to the driver and the reactive forces which are undesirable are triggered from the driver. The expansion of the clothoid curve with mathematical formula determines this target path. This is mainly motivated to associate the vehicle's current state (position, heading and curvature) to a future state of vehicle target. A typical clothoid curve is defined in the Eq. (3) given below:

$$(s) = \lambda s + \kappa 0 \tag{3}$$

where road's curvature is represented as j, the rate of change in the curvature is given by k, the distance down the spiral is given by s, and the initial curvature of road is denoted as j0. The data has been collected and object detected and classified. In this paper the object has been detected using RCNN (region-based Convolutional Neural network) and classified using AlexNet CNN architecture.

In an image various objects are distinguished by detecting and segmenting objects in R-CNN approach and the bounding box is drawn on the particular object. The method used for detecting and segmenting the object is Mask R-CNN. Bounding boxes are not only drawn by this Mask R-CNN and also pixels present in the bounding box are marked and classified even though is it is comprised of not comprised to the object and utilized for the object identification, but object's boundary is also marked and the key points are detected. Faster R-CNN forms the basis for R-CNN and its application is extended to the image segmentation field. Figure 1 illustrates the architecture of the network. RCNN process and Faster R-CNN process are similar and uses Region Proposal Network (RPN) for the feature extraction for the classification and tighten bounding boxes.

Every RoI region is quantified by extracting features, which is performed by R-CNN utilizes RoIPool and RoI feature's size problems varies with the scale are solved by max pooling. Though, Spatial information loss is caused by this process, creating the RoI with original image and misplacing the features extracted. For this problem resolving, Faster R-CNN's RoI pooling is substituted by ROI alignment (RoIAlign) through Mask R-CNN, mask branches are used consecutively for RoI Align result marking for the area of the object. The collected cases are split randomly into training sets and validation set in the process of training, the training data set established the model and it is tested contrary to validation set for ensuring the model's accuracy and stability. In R-CNN, the loss function value is $L = L_{class} + L_{box} + L_{mask}$ was reduced, and the highly appropriate design by the loss function minimization on the training data utilized as NN design. New data was predicted and analyzed by the trained model like validation set. The R-CNN's loss function is described in the below Eq. (4):

$$L = L_{class} + L_{box} + L_{mask} \tag{4}$$

where $L_{class} + L_{box}$ are recognized similarly in R-CNN, Eqs. (5) and (6) defines Lclass + Lbox:

$$L_{class} + L_{box} = \frac{1}{N_{cls}} \sum_i L_{cls}(p_i, p_i^*) + \frac{1}{N_{box}} \sum_i p_i^* L_1^{smooth}(t_i - t_i^*) \tag{5}$$

$$L_{cls}(\{(p_i, p_i^*)\}) = -p_i^* log p_i^* - (1 - p_i^* \log(1 - p_i^*)) \tag{6}$$

And the average binary cross-entropy loss is represented as Lmask and represented in Eq. (7):

$$L_{mask} = -\frac{1}{m^2} \sum_{1 \le i,j \le m} [y_{ij} logoy_{ij}^k + (1 - y_{ij})\log(1 - oy_{ij}^k)] \tag{7}$$

The quantitative evaluation of the trained R-CNN model's performance by mean average precision (mAP) as the lesion detection/segmentation's accuracy on the validation set provided by Eq. (8):

$$mAP = \frac{A \cap B}{A \cup B} = \frac{1}{N_T} \sum_{i=1}^{N_T} (\frac{N_i^{DR}}{N_i^D}) \tag{8}$$

Eight layers convolutional neural network is AlexNet having five convolutional layers and three full connection layers, performing maximum pooling operation after three convolutional layers. ReLU has used an activation function for AlexNet rather than the customary functions of sigmoid and tanh and so this AlexNet is varied from NN defined before. The activation function which is non-saturated is ReLU in the speed of training is not only enhanced effectively and also there is great management of disappearance and explosion of gradient's problem, and so the deep network is trained easily. The function of ReLU form is represented in Eq. (9).

$$\text{ReLU}(x) = \max(0, x) \tag{9}$$

The overfitting degree is reduced by the dropout in Alexnet, the model's training process stop the neurons with convinced probability, therefore the local node's dependence is decreased by improving the model's ability of generalization. Large convolutional kernels are introduced, in which the number of parameters is increased and in the process of extracting features the local features are lost; at the same instant, full connection layer parameter's proportion is comparative higher and on the outcomes has highly influenced by convolution part's extracted features, and because of it is specifically needed to improve the convolution parameter's proportion. Four convolutional layers, two de-convolutional layers and two full connection layers are present in this network. Features are extracted by 3×3 convolution kernel for avoiding few detailed features missed. At the same instance, down sampling uses the overlapped maximum pooling for decreasing the size of feature map.

De-convolutional layer present in the AlexNet is the main variation between this model and AlexNet. The number of feature maps is shortened by the operation of de-convolution and the full connection layer performs the output. Because of the number of feature maps simplification, highly reduces the full connection layer's input, the number of nodes is highly decreased in the full connection layer, the full connection layer's proportion of the parameter in the overall network is decreased. Over fitting degree model is reduced and the penalty factor to the loss function was increased for avoiding the too large or small parameters, for keeping the model comparatively simple. If 'J' is the loss function after regularization represented by Eq. (10)

$$\tilde{J}(\theta) = J(\theta) + \alpha\Omega(\theta) \tag{10}$$

as in Eq. (10). If there is no regularization, value of α is large $\alpha \in [0, +\infty)$, the penalty is greater and the regularization proportion is larger. The capability of generalization model is enhanced by L2 regularization in this paper and the weight of the model is weakened by this 0th approach represented in Eq. (11):

$$\tilde{J}(\omega) = J(\omega) + \frac{1}{2}\alpha\|\omega\|_2^2 \tag{11}$$

Normalization of data is performed by adding the batch normalization layer (BN) after the convolutional layer, which is not increasing the speed of converging network, gradient disappearance and explosion problems are resolved. The data is normalized by BN layer for the input sample's mean and variance calculation, γ and β are the two introduced parameters and the formula is represented in Eq. (12): in these the sample mean is denoted by μ, the sample variance is represented by σ, constant close to 0 is ε.

$$\begin{aligned} \hat{x}_i &= \frac{x_i - \mu}{\sqrt{\sigma^2 + \varepsilon}} \\ y_i &= \gamma\hat{x}_i + \beta = \mathrm{BN}_{\gamma,\beta}(x_i) \end{aligned} \tag{12}$$

4 Performance Analysis

Performance evaluation the algorithm proposed, uses the implementation of Traffic Information System utilizing the hardware environment of Intel (R) Core (TM) i5 (2.5 Hz) and the software environment of Windows 7, Visual Studio 2010 utilizing OpenGL and OpenCV Library. The novel AR system's design was focused in this research which influences the developing standard for vehicular communications, in the form of vehicle-to-vehicle and vehicle-to infrastructure traffic information systems. The objects appear before the car are successfully detected by this system and the information is reflected like speed, warning for driving or the direction of navigation within the windscreen for making the view simple with the eyes present in the road. The detectors are trained by the database which is gathered from the individual images. OpenCV Haar Training tools, the vehicles appeared to the front and back are classified with the resolution of 32 × 32 pixels to 64 × 64 pixels. Totally, three thousand images of vehicles are utilized in the process of training. Various experiments are conducted in various environments. Table 1 represents the vehicle recognition results. The Accuracy, precision, recall and F-1 score are the performance parameters used to measure the performance of the system and it produces better performance even in bad lighting conditions.

From the results of experiment, the cost of computation is reduced by this proposed method and the process of estimating the pose of camera is stabilized. For augmentation, real object attaches a virtual object with it, The superimposition of the real environment with virtual objects by the pose of the camera. The method of object tracking retains the right augmentation and the accurate calculation of pose estimation was proved. The performance analysis has been given in Table 1.

The above Figs. 2, 3, 4 and 5 shows comparative analysis of accuracy, precision, recall and F-1 score. It is observed that the proposed RCNN-ALENET CNN technique produced more accuracy than the existing techniques. The accuracy has been improved for proposed technique is as 98%. And in terms of precision, it is proved that the proposed RCNN-ALENET CNN technique produced higher precision than the existing techniques. By comparing precision analysis with existing techniques proposed technique has been improved by 94%. For recall it is noticed that the proposed RCNN-ALENET CNN technique produced maximum recall than the existing techniques. Recall has been improved as 92.3% for proposed technique when compared with existing technologies. The F-1 score of proposed work is enhanced

Table 1 Parametric analysis of AR-HUD with object detection

Parameters	CNN	ANN	SVM	AR-HUD RCNN-ALEXNET CNN
Accuracy	88	90	74	97.8
Precision	84	86	90	94
Recall	91.8	92	92.1	92.2
F1-Score	78.2	87.1	87.9	88.2

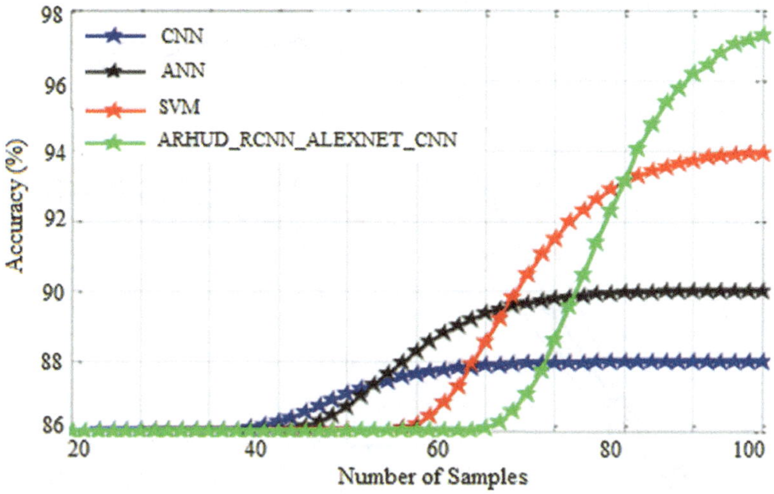

Fig. 2 Comparative analysis of accuracy

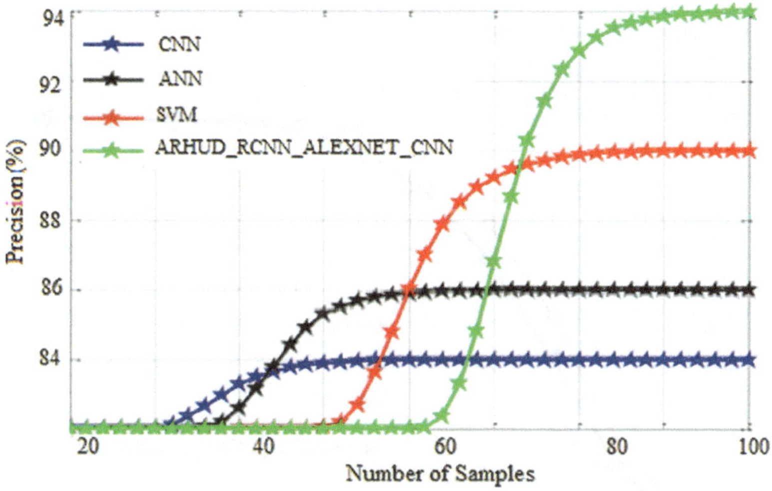

Fig. 3 Comparative analysis of precision

and it is discovered that the proposed RCNN-ALENET CNN technique achieves improved F1-Score of 86% than the existing techniques.

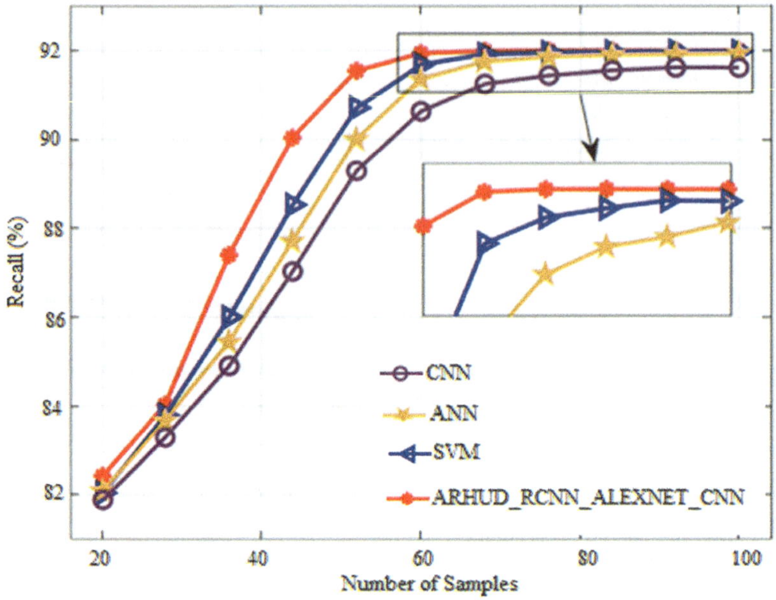

Fig. 4 Comparative analysis of recall

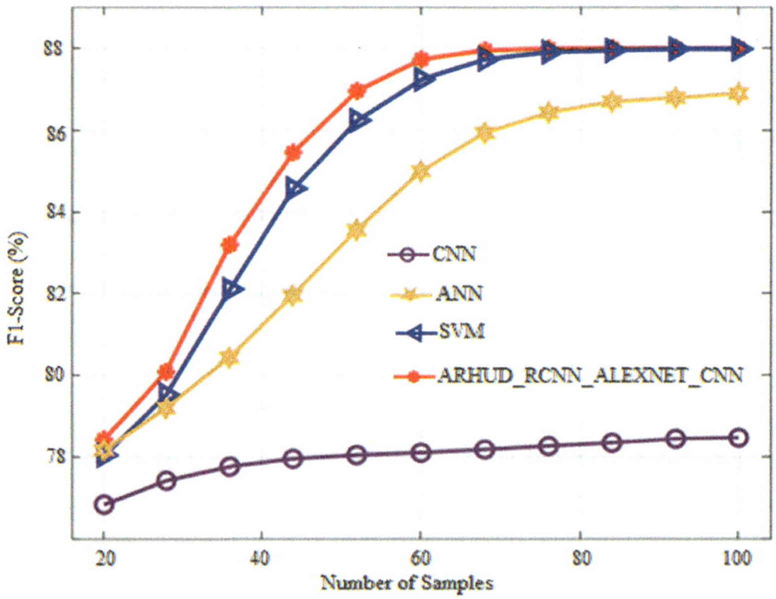

Fig. 5 Comparative analysis of F-1 score

5 Conclusion

A new dialog form was introduced between vehicle and the driver and it is facilitated by augmented Reality traffic information system; intelligent transportation systems were enhanced by the superimposition of traffic information, surrounded by the view of users and enables the view of drivers on the road. An In-vehicle AR-HUD system was proposed in this paper to provide the information of driving-safety in which the augmented virtual objects onto a real scene were superimposed. HMI design's context shows the efficiency, and HUD-based interfaces comparison for AVs are particularly applied and visual cues are provided the sensory and planning system of the vehicles. Exemplification of HMI's role is explored in provoking the trust sense and safety in AD approach, and this method is to surrender the vehicle control to AD from humans. The proposed technique shows human and machine interface (HMI) implementation for object detection and classification of moving objects. Here the object detection is done using RCNN and data classification is done using AlexNet CNN architecture. By simulation results, compared with existing technique, the proposed technique has been improved in terms of accuracy by 98%, precision by 94%, recall by 92.3% and F-1 score by 86%.

References

1. Wirza, R., Nazir, S., Khan, H.U., García-Magariño, I., Amin, R.: Augmented reality interface for complex anatomy learning in the central nervous system: a systematic review. J. Healthc. Eng. (2020)
2. Balters, S., Steinert, M.: Capturing emotion reactivity through physiology measurement as a foundation for affective engineering in engineering design science and engineering practices. J. Intell. Manuf. **28**(7), 1585–1607 (2017)
3. Rastogi, V.: An immersive vehicle-in-the-loop VR platform for evaluating human to autonomous vehicle interactions (2019)
4. Manawadu, U., Ishikawa, M., Kamezaki, M., Sugano, S.: Analysis of preference for autonomous driving under different traffic conditions using a driving simulator. JRM **27**(6), 660–670 (2015)
5. Jensen, L., Konradsen, F.: A review of the use of virtual reality head-mounted displays in education and training. Educ. Inf. Technol. **23**(4), 1515–1529 (2018)
6. Pratticò, F.G., Lamberti, F., Cannavò, A., Morra, L., Montuschi, P.: Comparing state-of-the-art and emerging augmented reality interfaces for autonomous vehicle-to-pedestrian communication. IEEE Trans. Veh. Technol. **70**(2), 1157–1168 (2021)
7. Calvi, A., D'Amico, F., Ferrante, C., Ciampoli, L.B.: Effectiveness of augmented reality warnings on driving behaviour whilst approaching pedestrian crossings: a driving simulator study. Accid. Anal. Prev. **147**, 105760 (2020)
8. Tabone, W., de Winter, J., Ackermann, C., Bärgman, J., Baumann, M., Deb, S., Stanton, N.A.: Vulnerable road users and the coming wave of automated vehicles: expert perspectives. Transp. Res. Interdiscip. Perspect. **9**, 100293 (2021)
9. Deb, S., Carruth, D.W., Fuad, M., Stanley, L.M.: Comparison of child and adult pedestrian perspectives of external features on autonomous vehicles using virtual reality experiment. In: International Conference on Applied Human Factors and Ergonomics, pp. 145–156 (2019)

10. Brown, B.: Assessment of human driver safety at Dilemma zones with automated vehicles through a virtual reality environment. In: 2018 systems and information engineering design symposium (SIEDS), pp. 185–190 (2018)
11. Ito, K., Hirose, M.: Immersive virtual reality environment to test interface of advanced driver assistance systems for elder driver. In: International Conference on Human-Computer Interaction, pp. 151–163 (2019)
12. Lin, H.: Using passenger elicitation for developing gesture design guidelines for 56 adjusting highly automated vehicle dynamics. In Companion Publication of the 2019 on Designing Interactive Systems Conference 2019 Companion, pp. 97–100 (2019)
13. Sun, Y., Li, H., Peng, W.: Robust validation platform of autonomous capability for commercial vehicles. In: WCX SAE World Congress Experience 2019. Technical Paper 2019-01-0686 (2019)
14. Leudet, J., Mikonos, T., Christophe, F., Mannisto, T.: Virtual environment for training autonomous vehicles. In: Annual Conference Towards Autonomous Robotic Systems, pp. 159–169 (2018)
15. Wang, Z., Liao, X., Wang, C., Oswald, D., Wu, G., Boriboonsomsin, K., Tiwari, P.: Driver behavior modeling using game engine and real vehicle: a learning-based approach. IEEE Trans. Intell. Veh. **5**(4), 738–749 (2020)
16. Hwu, T., Levy, M., Skorheim, S., Huber, D.: Matching representations of explainable artificial intelligence and eye gaze for human-machine interaction (2021). arXiv preprint. arXiv:2102.00179
17. Bram-Larbi, K.F., Charissis, V., Khan, S., Harrison, D.K., Drikakis, D.: Improving emergency vehicles' response times with the use of augmented reality and artificial intelligence. In: International Conference on Human-Computer Interaction, pp. 24–39. Springer, Cham (2020)
18. Rebensky, S., Carroll, M., Bennett, W., Hu, X.: Impact of heads-up displays on small unmanned aircraft system operator situation awareness and performance: a simulated study. Int. J. Hum. Comput. Interact. 1–13 (2021)
19. Safi, M., Chung, J., Pradhan, P.: Review of augmented reality in aerospace industry. Aircr. Eng. Aerosp. Technol. (2019)
20. Yontem, A.O., Li, K., Chu, D., Meijering, V., Skrypchuk, L.: Prospective immersive human-machine interface for future vehicles: multiple zones turn the full windscreen into a head-up display. IEEE Veh. Technol. Mag. **16**(1), 83–92 (2020)
21. Bremers, A.W., Yöntem, A.Ö., Li, K., Chu, D., Meijering, V., Janssen, C.P.: Perception of perspective in augmented reality head-up displays. Int. J. Hum. Comput. Stud. **155**, 102693 (2021)
22. Newman, B., Carlberg, K., Desai, R.: Optimal assistance for object-rearrangement tasks in augmented reality (2020). arXiv preprint. arXiv:2010.07358
23. Jaiswal, S., Amin, R., Sopanen, J., Mikkola, A.: Real-time multibody model-based heads-up display unit of a tractor. IEEE Access **9**, 57645–57657 (2021)
24. Xiong, J., Bi, R., Zhao, M., Guo, J., Yang, Q.: Edge-assisted privacy-preserving raw data sharing framework for connected autonomous vehicles. IEEE Wirel. Commun. **27**(3), 24–30 (2020)
25. Pranav, K.B., Manikandan, J.: Design and evaluation of a real-time pedestrian detection system for autonomous vehicles. In: 2020 Zooming Innovation in Consumer Technologies Conference (ZINC), pp. 155–159. IEEE (2020)
26. Aladem, M., Rawashdeh, S.A.: A single-stream segmentation and depth prediction CNN for autonomous driving. IEEE Intell. Syst. (2020)
27. Cai, P., Sun, Y., Wang, H., Liu, M.: VTGNet: a vision-based trajectory generation network for autonomous vehicles in urban environments. IEEE Trans. Intell. Veh. (2020)

Special Sensors for Autonomous Navigation Systems in Crops Investigation System

Kuldeep Singh Kaswan, Jagjit Singh Dhatterwal, Anupam Baliyan, and Vishal Jain

Abstract Mobile robots functioning in farmlands have been an important focus for scientists. The fast improvement in communication, sensing and computer technology has given considerable advances to Robot navigation guidance techniques in agriculture. Automatic autonomous robots minimize work expenses, avoid dangerous activities from being carried out by people and give farmers timely and accurate data to help management choices. Appropriate methods for sensing, mapping, localization, trajectory planning, and preventing obstacles are designed through research into robot sensor technologies in agricultural contexts. The Navigation Algorithms must use visual information to determine an acceptable course, execute a selection and navigate appropriately without collisions in its environment. A summary of sensor technology for autonomous prototype vehicles is presented and discussed in this chapter. Navigating sensors, computer methods, and navigation management approaches are the main aspects. Crucial procedures include selecting, coordinating, and combining the appropriate Sensors to provide essential robotics navigational knowledge. For function extraction, processing of data and fusing computationally efficiently are utilized. The steering controllers give the correct steering motion to operate automated vehicles for autonomous navigation. Mobile robots are still an open topic in outside contexts such as in agriculture. To address the challenges posed by the climatic conditions, dynamic surroundings, unforeseen obstructions, terrain variations, and vegetation, it is necessary to provide effective and powerful protective and actuators technologies for mobility farming robotics. In this chapter, we will

K. S. Kaswan
School of Computing Science and Engineering, Galgotias University, Greater Noida, India

J. S. Dhatterwal (✉)
Department of Computer Science & Applications, PDM University, Bahadurgarh, Haryana, India

A. Baliyan
Chitkara University Institute of Engineering & Technology, Chitkara University, Rajpura, Punjab, India

V. Jain
Department of Computer Science and Engineering, School of Engineering and Technology, Sharda University, Greater Noida, India

A. E. Hassanien et al. (eds.), *Virtual and Augmented Reality for Automobile Industry: Innovation Vision and Applications*, Studies in Systems, Decision and Control 412, https://doi.org/10.1007/978-3-030-94102-4_4

discuss about special sensor keep monitoring through GPS system requirement of crops and to improve and fine growth of quality seeds.

Keywords Agriculture · Autonomous · Control · Laser · Localization · Mapping · Mobile robot · Navigation · Sensing · Vision

1 Introduction

In the diverse interior and exterior farming contexts mobile robot systems have been deployed. In many irrigations scheduling, these robots play an important role, since they minimize human work and improve operating safety. In various agricultural works, for example, planting, sprinkling, fertilize, grow, harvesting, trimming, wetting and inspection, the necessity for autonomous navigation of mobile robots has already been recognized see in Table 1. These machines should have enough cognition to be sensitive to certain duties for extended periods [1]. Data source is https://www.kaggle.com/uciml/wall-followingrobot/version/1?select=sensor_readings_4.csv (Fig. 1).

The designing of moving robots, such as smart agriculture, in outside situations is still a challenge. Agricultural navigation creates problems because of changes in climate and differences in topography and vegetation character. These environmental characteristics must be addressed, and effective, robust sensing and control systems

Table 1 Robot movement in agricultural area

Movement 1	Movement 2	Movement 3	Movement 4	Movement position
1.687	0.445	2.332	0.429	Slight-right-turn
1.687	0.449	2.332	0.429	Slight-right-turn
1.687	0.449	2.334	0.429	Slight-right-turn
1.687	0.449	2.334	0.429	Slight-right-turn
1.344	0.496	2.843	0.692	Move-forward
1.33	0.498	2.836	0.701	Move-forward
1.32	0.499	2.83	0.71	Move-forward
1.31	0.501	2.824	0.719	Move-forward
0.898	0.638	2.524	1.137	Sharp-right-turn
0.881	0.645	2.511	1.155	Sharp-right-turn
0.864	0.652	2.499	1.172	Sharp-right-turn
0.847	0.658	2.487	1.193	Sharp-right-turn
1.821	0.371	2.158	0.836	Slight-right-turn
1.817	0.363	2.239	0.832	Slight-right-turn
1.816	0.357	2.24	0.831	Slight-right-turn
1.814	0.357	2.244	0.833	Slight-right-turn

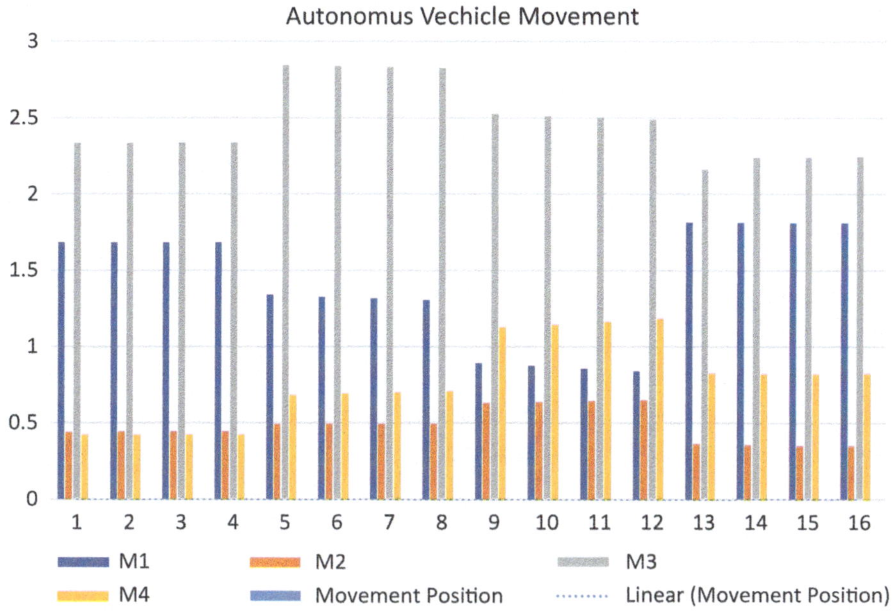

Fig. 1 Autonomous vehicle movement

must be designed and designed. By detecting the surroundings and applying suitable control algorithms for specific tasks the autonomously of the mobile robot is achieved. Before the robot accomplishes the mission, the robot coordination and prototyping protocols must be designed and optimized. To get good results, a full understanding and limits of the robotic facilities, environmental features, and task needs are required [2].

Prior examinations have shown several elements of autonomous mobile robotic navigating in farming contexts. This chapter analyzed a range of common farm settings and scenarios and proposed appropriate sensor designs to fulfill the needs for diverse activities of agricultural robots. A review study on farming automation systems examined the subjects of sensing and perceiving, cognition and understanding, evaluation of information, and task planning and implementation [3]. An overview of agricultural vehicle guiding technology was provided in the chapter. The writers focused on sensors, computing techniques, navigational planners, and monitoring systems. This chapter is intended to give an overview of new advances in farming automated navigation. In the last 15 years, this chapter is focused on global research in autonomous navigation. Mobile agricultural robot's advanced automation system includes Movable robot's browser sensing, navigational management strategies, and computer technology. The basic characteristics of the mobility independently guidance system is outlined in Fig. 2.

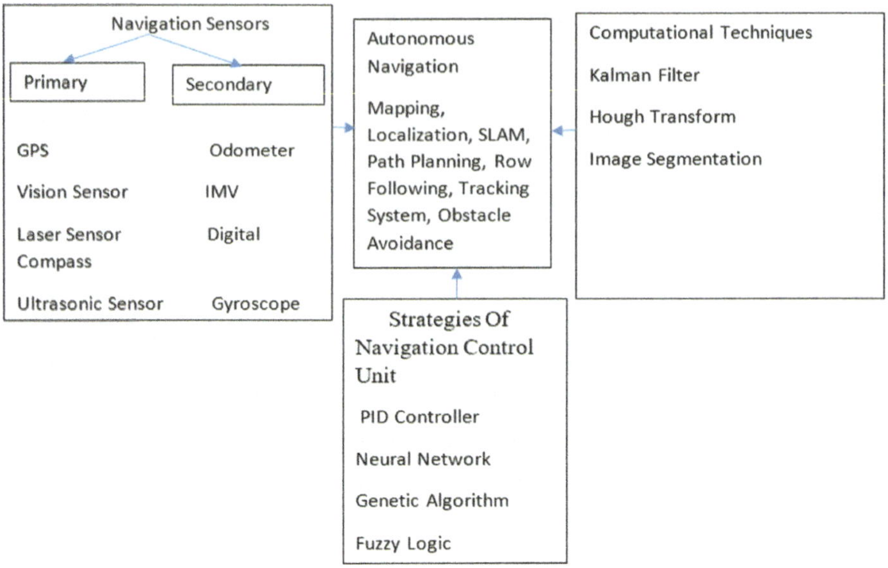

Fig. 2 Fundamental components of the mobility farming robots autonomously navigation

2 Purpose of Chapter Description

The following chapter has been organized. This section outlines some of the problems and benefits inherent in the development of autonomous farming sailing and the principal issues linked to the work of autonomously agricultural sailing. The chapter next examines the literature based on the most important autonomously features of the robotics navigating technology the study is divided into three sections:

- Mobile sensors of navigating the robot.
- Methods of computing.
- Strategies for navigational control.

Differentiating autonomous navigation in different crops in fields and tree line rows in the garden will be a key issue in these three categories. Finally, a summary is given of the important findings of the literature examined.

3 Mobile Robot Navigation in Farmland Environments

There are numerous aspects in agriculture such as vegetation, trees, weeds, soils, items, and sites of interest. This variety poses certain problems and complexity for the navigating of the mobile robot [4]. The farmed area is big, for example, and the textures of the land are generally uneven. The data obtained from mobile robot sensors may impact climatic conditions, such as rainfall, dirt, mist, and sunshine. The plant hue, which may change from location to place, may alter throughout different phases of development. The farming setting, however, is simpler [5]. For instance, the majority of the identical kind plant parts are sown in parallel directions, with approximately similar spacing between the lines. The existing landmarks may be employed in localization and navigating algorithms as permanent landmarks.

Special problems with mobile robots are autonomously navigating. The aim of navigating is to navigate the robot in diverse settings safely and independently. The navigational capability of the robot is based on cognitive technologies and sophisticated sensing systems [6]. The robot must initially be able to identify, detect and model the environment. Mobile robot movement in a farm environment requires consideration of the positioning, surroundings, and barriers of the mobile robot. Incidental impediments such as deceased and living creatures, falling trees and bushes, and closed doors should not have been prevented by the mobile robot. A mechanism to identify, map, estimate positioning and browse any such device should be in place [7].

Many academics have developed automated robots that might investigate their system and improve an environmental map. To improve the resilience of the map the integration of several sensors has been used. The robot can simply detect its position and size each time when the map of the surroundings is correct. Localization and mapping are interconnected processes, since a good location is needed to create a good map, and a precise map is vital to be well located [8]. The approach used by driverless driving to construct a map in the unknown area is simultaneously location and mapping (SLAM), while simultaneously using this mapping to locate their present location. For automated operation, route planners are necessary to identify the best way from start to finish so that no barrier congestion occurs [9]. Since many plants are planted in rows, the development of a Row predictive algorithm is essential to allow mobility robotics to move the rows see in Fig. 3. The difficulty lies in the line identification procedure to discover precise characteristics stable under various

Fig. 3 Robot navigation in the agricultural field

situations in the surroundings. The row identification procedure has some challenges, such as incomplete lines, misplaced Row predictive algorithms that are essential to allow mobility robotics to move the rows [10].

4 Navigation Sensors Improve the Quality of Crops

In recent years, sensors have been used in driver assistance systems fast. Navigator sensors offer information on vehicle conditions and objects in the natural atmosphere (position, orientation, speed, etc.). Some guide sensors offer absolute location information while others offer relative positioning only [11]. In autonomous mobile robot technologies, various sensors are utilized for principal sensing's, such as Global Positioning System (GPS), vision, and Thermal Imaging Analyzer. The literature also reports that ultrasonic sensors and RFIDs are key sensors, but they are less prevalent. As a supplement to main sensing applications is generally employed as additional sensors other sensors such as the odometer, inertial unit (IMU), motion detector, and accelerometer [12] see in Fig. 4.

4.1 Surveillance Navigation Technology

The efficiency of sensor technology and their capacity to offer enormous effort to develop steering control signals for mobile agriculture robots have been frequently

Fig. 4 Navigation sensor

utilized in robotic navigation. Motion control functions including localization, map generation, autonomic navigation, training, inspections, surveillance, and obstacle avoidance have become much more widespread in outdoor smart agriculture. The effect of different ambient lighting levels in outdoor locations is a major drawback with vision sensors [13].

Various sight detectors have been used to identify parameters for grain boundaries. The placement and harmonization of the for example different crops concerning the vehicles and the margins along harvested cereals are detected. They were particularly concerned with the development of several approaches of picture fragmentation to derive advice on Grain lines [14]. The visions of the guideline were investigated generally responsive to weeds sensory disturbance The authors' method utilized the picture to track the time consumption and row matching for the optimum line for the cropping row the authors created this technique. Okamoto et al. studied an automated row tracking system for a wedding grower using the color CCD camera (2002). This method used the pictures of the crop line to calculate the offset here between the grower and the desired crop line [15]. The offset predictions enhanced the exactness of the subsequent row. A single monochromatic camera described a visual guide system for grain harvesters. The guiding algorithm is evaluated on the side location of the cutting edge and was able to arrange crop rows properly. A row classification method was suggested to acquire direction for steering the tractor with a single-chrome CCD camera. For unique emotional guidance, the algorithm was satisfactory in terms of accuracy [16]. It has presented a camera technique for solving the challenge of autonomous navigation. Features removed from the photos that the robot needs to travel to map the plantation rows. An autonomous field inspection vehicle was created and utilizing a camera that navigated among two different crops. The position of the related vehicle was established by the classification and segmentation of the picture and by the extraction of geometric lines for the crop rows. Researchers reported that spectrum filters are utilized to improve Detection after implementation of depth perception row crop guidance. The detection of planets in near-infrared (NIR) imaging produced good results in these systems [17].

Investigation on the aerial mapping of tree rows of mobile robots in orchards and groves is rather rare. The challenge of locating a feasible path from rows resides in discreet habitat restoration which disrupts a row's visual consistency and makes the segmentation process more challenging. A variety of automated vision-based browsing devices for the tree line were created. These systems employed several methods of categorization and segmentation of data to extract important visual feedback and to optimize classifications. Traditionally, line detecting techniques are employed to find guiding routes [18].

Stereoscopic view technologies are also investigated to give 3D field pictures by merging concurrently two field photos in food applications for autonomous navigation. The background between the two facilities and the quality of length measuring diminishes as the depth rises, limit precise and actual measured readings. Stereo-based crop row detecting systems have been developed. The results show that the amount of information and precision required to create the 3D fields, location, and navigation systems may be achieved using stereo perceiving [19].

4.2 Frame Land Operation Through Global Positioning System (GPS)

Management technology based on the global positioning system has been extensively employed in various farm operations. GPS guiding systems offer absolute location readings for various outdoor applications utilized in navigating the mobile robot [20]. Additional devices have been introduced by various universities to the typical GPS tracking technology to improve its effectiveness. The technologies presently developed include the Global Positional Differential System (GPS) and the Global Positioning System (RT-GPS), which are kinematics real-time technologies. Need certain distances from the farming robot receiver to be placed inside the base of the GPS. An investigation has shown that the RTK-GPS has been the only location sensor for the agricultural automated steering column [21]. Whether the GPS type is the one-position sensor for the self-sufficient navigation of mobile robots, the GPS has several limitations. A GPS typically provides more precise navigational information in conjunction with other sensors. Examples of the combination of RTK-GPS and IMU are available in the following. The research was developed utilizing autonomous agricultural vehicles RTK-GPS and fiber optic gyroscope (FOG) sensors suggested the development of an automated weeds management system using an RTK-GPS tilt sensor [22].

The most prevalent difficulties with GPS navigation are blocking satellite line-of-view, multi-path difficulties, and other RF sources. GPS cannot be used in certain locations or browsing vergers because farming robotics travel often under the dense forest blocking communication satellites to the Navigation system. Therefore, several researchers create intelligent electronic robot navigation systems without Global positioning as the fundamental agronomic navigational technology [23] see in Fig. 5.

Fig. 5 Farmland operation through GPS

4.3 Laser Scanning and Navigating Dependent on Perception

The cameras and laser scanning can be employed as the primary sensor for autonomous driving in some semi-natural situations, such as fields or verging areas. The connectivity of equipment and a laser scanner offers robust assistance for the independent robot navigation system by increasing the capability of detecting objects. The work of an autonomous guide system for the control and rotational motion, based on computer vision and laser radar. The tractor was autonomously driven by the guiding system on straight and curving tracks. SLAM-based systems have been developed using a combining of laser scanning and cameras utilized Extended Kalman Filter (EKF), However, using the Extended Information Filter (EIF), it was found more appropriate to use EIF as products for real-world applications in the advanced period [24].

4.4 Other Sensors (Ultrasonic and RFID)

In crop recognizing robot manipulator navigation in agricultural contexts, monitoring systems have been employed. Its usage is nevertheless limited in exterior applications as it requires that the objects are orthogonal to the sensor to adequately reflect the ultrasonic echoes. To recognize the cornfield row to maneuver the robot in an agricultural setting, a sonar-based crop row mapping approach. Ultrasonic sensors for plant recognition are employed for navigational purposes. You used echo correlations from several directions to improve plant reconnaissance effectiveness. In the study carried out, monitoring systems were used to compare the actual location between a tractor and a tree dock for the orchard's tractor navigation. A new location technique for RFID mapping systems has been developed in the chapter given. For identification of the plants and bins, specialized passive RFID tags were used, and an RFID reader was placed on the gathering equipment [25].

4.5 Laser Scanner-Based Navigation

Several scientific contributions were made by navigating using laser scanners. This is mostly due to high precision and broad field of view advantages for thermal cameras. One of the most common outdoor application equipment is laser scanning [26]. By monitoring the flight duration of charged particles, the comparative proximity of items in the neighboring area is estimated. The ability to provide robust data for image classification and localization is a key benefit for laser sensors over visual systems. This allows the robot to function more dependably under varied circumstances of climate and ambient lighting [27]. The use of laser sensors to identify crop rows to derive guidelines has been described. Various techniques were used for line detecting and information fusion. As navigational sensors, laser technology is commonly linked to other sensors utilizing adequate information fusion techniques. The automated row guiding system for tractors was studied, which used a laser sensor that could identify the height and location of crop lines. It has built an autonomous robot tractor to collect geographical information on cultivation and output using a scanning electron microscope. For the location, mapping, and navigation for autonomous agricultural robot utilize 3D LIDAR flora identity and soil optimization sensors. The 2D laser data were concurrently utilized to track the development and independent navigation using artificial landmarks in the work performed [28].

The prospective usage of tree row detecting laser sensors in orchards and forests have also been studied by researchers. A laser scanner may be used for detecting distinct tree row components (e.g., trunk, stem, and canopy), and normally just plants canopy for crop cultivation can be identified. To determine a thick canopy in the tree rows, employed a scanning electron microscope [29]. However, the employment of reflexive tapes to identify line endings has decreased processing time and improved row identification. A technique for the detection of tree row trunk and/or canopies of branches was provided. Hough transformation was then used to extract point and line characteristics for navigation between the rows of the agricultural vehicle. A radiation source was utilized for performing SLAM algorithms using the treetops as feature points in the study reports [30].

In other experiments, photogrammetry was used to estimate the distances between trees, and barriers of trees in the vertebrae, and then the line detection approach was employed to identify tree rows. The orchards rows are recognized using Hough transformation. In estimating the position and size of farming mobile robots using various data fusion techniques the separation distance of the trees recorded by the scanning laser were employed [31].

5 Advanced Computational Techniques

To extract essential environmental information, complex algorithms and methodologies are required to drive cars independently. Various computing approaches for object detection fusion are utilized to give information necessary for autonomous vehicles by agricultural vehicles. Kalman filter, Hough transformation, and picture classification are the most often utilized computing algorithms in the literature [32].

5.1 Hough Transform

Hough transformation is a method used for the extraction of images, electronic imaging transformation, and computerized intelligence. The features of a specific shape are separated in an image. The traditional Hough transformation was based on the identification of lines in the picture, but subsequently the positions and inclination of specific kinds, usually circles or ellipses, were enhanced. Row detection is mostly linked to independent navigating. Since most cultures and trees are maintained in rows (typically in straight lines), Hough is used to identifying most image processing techniques. In several experiments, Proposed Method was effective in the identification of crop or tree rows utilizing visual or/and laser scanners as a navigational sensor. Utilizing Hough to recognize tree rows and to build a road connecting two rows. It proposed a way of recognizing plant rows strong based on the process of Hough, which may direct agricultural machinery. The innovation of the algorithms was that they modeled a plant row instead of just a line with a pyramid shape [33].

5.2 *Information Fusion in Sensor Provide by Kalman Filter*

A strong mathematical approach for real-time information fusion in many sensors is provided by Kalman Filter (KF). To analytically evaluate the stability of the process each instant, the numerous observations that are collected throughout time with random fluctuation (bruising) are merged. Not only can this approach assess the current or prior condition, but it can also forecast the future status. The KF deluxe edition is for processes represented by the linear stochastic partial differential equation. Mobile robot services are typically non-linear. This filter has thus been expanded. Extended Kalman (EKF) Filter was changed to tackle the difficult nonlinear challenging problem [34].

This method predicts the orientation (orientation) of vehicle models using KF once suitable models have been developed for the device and the instruments. The data gathered from various sensors are largely utilized in this chapter. Some chapters have suggested integrating the GPS into other sensors that use KF to enhance estimates of location accuracy. KF is commonly used for connecting laser data with information from other sensing devices for robot placement in plantations. The EKF algorithm was utilized for the localization of a mobile robot in an orchard setting while using the laser, odometer, and IMU scanner to enhance position estimates. To increase its performance, EKF frequently combines several approaches of optimization with control schemes. The conduct of automated driving with machines view, laser radar, IMU, and communication module, established an upgraded KF logic for ultrasonic sensors [35].

With various forms of filters mentioned in the literature, SLAM issues may be effectively handled. EKF might be regarded as the most often utilized SLAM algorithm. However, the time required and computer demands of EKF-SLAM are disadvantageous. With the number of locations and functions on the map, the sophistication of EKF-SLAM is increasing. EKF was created to implement the SLAM issue in some research [36]. Kalman (UKF) unscented filtering can be used to support the SLAM method as an alternative filter that produces superior performances than the EKF concerning the process and observational modeling nonlinearity. On the other perspective, due to its linear calculation costs, the extended Information filter (EIF) increases the SLAM algorithm's processing speed. The EIF is more suitable to be used in an actual moment the difference between the EKF and the EIF is that information is distinctive. To describe the uncertainties of the SLAM, the EIF employs the inverted covariance, known as the information matrix [37]. The SLAM challenge in the olive tribal detection through vision and laser sensors. The EIF-SLAM technique has been optimized to enhance processing and estimate time. The SLAM difficulty is also solved with the Particle Filter (PF). In contrast to the EKF, the PF is not bound to Gaussian processes and manages nonlinearities related to the evaluation process better, although its real-time delivery is still constrained. A self-localization method consisting of a 2D laser sensor and a particulate filter for sensor uncertainty. The use of three derivative-free filtering for robotic location and surfing in a vineyard [38]. To adjust the predicted posture by filtering the solution employs the rows of the tree as measures. They discovered that the highly speculative filters are more adaptable for system modifications and representations of the measurement items.

5.3 Collection of Digital Image Segmentation Through Autonomous Vehicle

Machine intelligence segmented refers to the technique used to divide a digitized photograph into several regions or pixels gathering. Usually, items and boundaries in photos are detected in autonomous vehicles (lines, curves, etc.). The image division techniques have been developed in agricultural contexts to divide items into distinct classes [39] (crops, backgrounds, weeds, trees, etc.) to extract instructions for the row following.

In the literature, several approaches have been explored for extracting guiding information from the picture segmentation approach. The row classification technique based on K means to drive a tractor in straighter and curving lines illustrates that a boundary in the range of values is used using the k-means clustering method (ROI) [40]. The distribution separation and edge identification of crop rows were explored. The chapter utilized to extract geometric lines matching to crop row using a mixture of histogram-based approach, thresholds functional, and morphology photography functions to determine the relative location of vehicles. An approach to detect parallel rows with a mixture from edge-based and the Hough transformation has been proposed by Ericson and Astrand. They introduced a new approach on the edge of photographs to identify lines and additional rectangles. A unique technique, called Fourier imaging rotations and projection algorithm, for matured wheat cutting and uncut edge detecting lines [41]. Two essential operating conditions, the heading angle and the sides are calculated using this technique. A method to identify the required path between the tree rows by combining boundary and image segmentation. Perceptual color grouping and morphology machine learning were utilized in the work to achieve segmentation pathways, The approach utilized to get the best designing the application was then least-square curve fitting. The regional segmentation process and dual-Hough transformation for cultivation row identification. For the classification and segmentation of farming sceneries to the navigation issue, utilized JSEC Algorithm with Artificial Neural Networks (ANN) [42].

The JSEG segmentation is divided into three stages: quantification of color space hit rate areas and the merger of comparable color areas.

6 Strategies of Navigation Control Unit

A tough problem is the designing of the control unit for agricultural vehicles. Mobile farming robots generally work on several types of terrain, uniformly and unpredictably, or changeable topography. Steering controllers need to be able to regulate the steered commands for automated driving in reaction to mobile robot variations, travel speeds, terrain, and other elements that impact steering dynamics. Various control techniques, such as PID, Neural Network, Genetic Algorithms (GA), and Fuzzy Logic (FL) have been found in the literature. Variable structure barriers have been developed. Significant improvements have been made that can direct autonomous mobile robots to a target point in known or unfamiliar settings [43].

6.1 Neural Network (NN) and/or Genetic Algorithm (GA)

NN and/or GA were used to control the mobility of robot manipulators in agricultural production applications. To reflect the link between vehicle movement outputs and inputs on sloping ground, an NN simulation tool rather than a dynamic or film model. For estimation of vehicle behavior, designed a simulation tool for slope terrain. The NN model was trained via a training approach coupled with the GA and back spread algorithm. A specified track directed the tractor effectively. A GA route planer for agricultural machines. GA has been chosen to provide the vehicle with an optimum path to traverse an area fully without any known impediments [44].

6.2 Robotic Navigational Based on Proportional-Integral-Derivative (PID)

In several investigations of mobile robots guiding systems in agricultural contexts, a proportional integral derivative (PID) controller has been employed. A method to autonomous navigations in the agricultural context using proportional-integral (PI) controls were presented [45]. The findings showed that the robot had an appropriate behavior in real planting. A geomagnetic position sensor guided PID steering controller for a farm tractor. Its closed-loop transfer function has been experimentally obtained and test results demonstrate the successful automated guiding of the controller. A PID controller, which reduced track errors and steeds the tractor down the alleyway of a citrus orchard using vision system information and laser radar, has been created. Another example for controlling a farmer's location using the PID controller is the row recognition system information. PID was also suggested for the steering of the intra-row weed management unmanned hoeing system [46].

6.3 Robotic System Control by Fuzzy Logic (FL) Technology

FL control is ideally suited to control a robotic system, as it can deduce even amid uncertainties. It also addresses complicated and non-linear activities efficiently. In the autonomous vehicles of mobile robots in agriculture, the FL microcontroller has been employed. It described a navigational technique using sound mapping of the crop lines and FL control in a farming setting to manage a mobile robot on wheels [47]. The application of the FL controller to build a machine vision navigation system to direct a farmland harvester combine. Results showed an exact cutting-edge identification of the algorithm. In the study to combine navigational data from differentiated GPS and sight engines. In this investigation, the FL model was used. The unmanned navigation of the robot tractors under varied soil conditions describes the introduction of a robotic steering controller in safe obstacles [48].

Various adaptive fuzzy methodologies are employed, which are flexible and adaptable to farming settings in the outside area, such Neuro-Fuzzy and Genetic-Fuzzy. The combines of NN and FL are neuro fuzzy. A NN helps to learn and adapt, while FL helps to develop rules and make decisions. The neuro-fuzzy method to the autonomous mobile robot. To adjust the fluid rules and the membership functions dynamically, an educational algorithm based on NN was created. A neuro-fuzzy-based system was presented for the responsive navigation of a mobile robot with comportment-based control in the investigation. The technique presented employed optimum NN training with discontinuous samples [49].

GA algorithms have been utilized in recent years to automatically learn the flowing control parameters and affiliate operations for independent mobile robot navigation. The FL controller was utilized to operate the orchard velocity spraying independently. The GA was used to optimize the created fuzzy logic controller. A new genetic approach to train and adjust online the advanced software navigation system. This technique might be utilized in unstructured and changing settings by automated vehicle communication [50].

7 Smart Agricultural Monitoring to Optimize Farming Productivity

Improved agricultural systems are clearly distinguished from any IoT-enabled solutions. Agriculture is also one of the highest progressive rates for Digital transformation, a key industry inside the world economy. Improvements in the business are robust: the world population is projected to exceed 10 billion by 2050.

Given these expectations, it is inconceivable to ignore the necessity of agricultural surveillance. Of course, those firms which manage, by using agricultural surveillance systems, to meet the rising demands for sustainable food items will acquire a clear competitiveness lead.

- **Regular maintenance of equipment**

Machinery is a major factor for agriculture as an enterprise. Even on a timetable, management plans take time and have an influence on the budget, but the unpredictable aspect is nonetheless not eliminated. When a piece of machinery falls out of order by mistake, it generally leads to unplanned interruptions.

- **Water forecasts are correct**

Water is necessary for cultivated plants, although the quantities vary based on the soil moisture. Farmers must travel the field and do frequent manual tests to monitor these levels, alternately using intelligent sensing equipment, which is considerably more precise, practical and time-effectiveness in Fig. 6.

- **Waste disposal and overhead costs**

Under watering or irrigating of the plants might happen if precise soil humidity data are not collected. Poorly hydrated plants are dry and fragile, yet potassium deficiency causes contaminated water and implies unforeseen water costs.

- **Assessment of accurate seeding timings**

Depending on a range of environmental conditions, each plant has an ideal planting period. However, without proper data, it is frequently impossible to predict correctly this time.

Fig. 6 Checking water level in soil

- **Temperature and humidity measurement**

Soil temperature and humidity are essential variables for agricultural producers to gather for estimation of crop conditions and act accordingly. Unfortunately, without IoT agricultural monitoring devices, it is typically difficult to evaluate them properly.

- **Control of the pestilence**

Another issue for farmers is the management of pests, the locations, activity and behavioral patterns. Unsurprisingly, without IoT-based pest control devices this problem is also extremely impossible to address.

8 Smart Agriculture Monitoring Solutions

IBM forecasts that by using IoT, farmers will achieve a 70% output rate by the end of 2050, thus the future is overall bright. In certain respects, IoT has much to offer to alleviate difficulties routinely faced by farmers.

Agritech is a prosperous sector, which now helps farmers to face their everyday problems with a wide range of intelligent agricultural technologies. Planting, irrigation, cultivation and pest control—the surveillance of the field in agriculture collects a variety of metrics for farmers to successfully handle these activities.

- **Monitoring of soil condition**

The soil is an essential indication that helps farmers to determine the optimum plantation and harvest period. Farmers are immediately notified to the moisture content and saltiness with IoT devices that perform soil condition monitoring. Additional metrics have included the temperature of the soil and of the air, and the proper estimation allows farmers to schedule watering schedules and to know when to anticipate pests.

Soil condition control requires a mix of hardware and the software to work and notify users to major patterns in income time. CropX—an ag-tech network for remote agricultural monitoring is an example of such a solution. It utilizes intelligent agricultural sensors to gather data and a data processing and storage cloud infrastructure to supply information in a legible manner to a user PC or to a computer monitor in Fig. 7.

- **Monitoring of the weather**

In farming, weather forecasting is one of the most common applications of IoT. In agricultural cultivation, the yields depend largely on the naturally unpredictable climate. Weather monitoring systems on site (such as weather stations), notify farmers to changing circumstances—temperature, rainfall, moisture, solar irradiation, and wind velocity. The use of remote monitoring technologies in agriculture allows farmers to provide effective weather reports straight to their computers and cellphones, thus allowing them to take prompt action, is an important example. Weather control platforms include Pyknon, all METEO and Smart Component in Fig. 8.

Fig. 7 Monitoring sensor soil condition

Fig. 8 Sensor of weather monitoring

- **Automation systems for greenhouses**

Incessant care and management are needed in a delicate and sensitive greenhouse ecosystem. Smart greenhouse automation solutions such as Grow link, Far Mapp and GreenIQ show remote sensing applications in agriculture. They assist preservation and management of light, humidity, CO_2, and temperatures concentrations in the optimum microclimate circumstances. The effectiveness of greenhouse farming is maximized by immediate warnings and enhanced administration skills in Fig. 9.

Fig. 9 Sensor of green house

- **Systems for crop surveillance**

As crops develop and mature, so much may go completely mistaken infections, pests or unfavorable health situations may inflict irreparable harm long before farmers realize. The smart imaging technique, which is used to crop monitoring, captures crop status data (weather, moisture, health factors) and allows farmers to undertake prompt action should something go wrong.

Furthermore, devices like Semios and Arable assist to determine when crops are ready to be harvesting in Fig. 10.

- **Digital pesticides**

Some of the greatest crop growers often suffer plague infestations. While it may be difficult to know when pesticides arrive, it is generally impossible for them to identify their activities and area without regular field excursions. Smart farm surveillance systems address these issues and assist in the allocation of the appropriate quantity of pesticides necessary to remove pests in each specific situation.

Fig. 10 Crop surveillance system

IoT pest detection devices such as Strider count mosquitoes and locate them with a straight insect and pest sensors probe on the ground in real time. Similar technologies are available for IoT-based pest management by Ag-tech firms like Fielding and DTN.

- **Systems for animal surveillance**

Agriculture monitoring systems also acquire broader applicability in animal production, apart from agricultural and meteorological surveillance. The combination of advanced IoT hardware, such as remote monitoring equipment, and state-of-the-art IoT software, helps secure and defend animals using ag-tech products, such as Important to first establish. SCR is another farm video conferencing firm employing cow neck collar to track cow health, position, and activities. Fern-sensing and modern software program in agriculture provide information about cow feeding and the condition of the complete cattle.

9 Conclusions

From the need for labor to complete most farming production to the use of fully automatic farming machinery, farming technologies have traditionally progressed. It is suggested from the literature review that a more robust guiding mechanism is provided by a robot manipulator with multi-sensor combinations (primary and secondary sensors). The fundamental idea underlying ultrasonic sensors is to achieve better navigational, localization, and location assessments of mobile robots through the combination of diverse sensor inputs. On the other side, the use of multiple sensors increases system design costs and complexity. GPS may not be effective for a covered canopy in a garden from the prior experiments. The scanners for cameras and lasers may be seen as the best sensors that are utilized for navigation as key sensors. They can be used alone or with other auxiliary sensors like speedometer, IMU, compasses, and gyroscope. They may be used independently.

Hough transformation may be used to map the surroundings and for row detection with an appropriate picture segmentation method. EKF may be seen as the most successful option for sensor data synthesis for location and SLAM, which allows for an exact assessment of robot navigation positioning and guidance. Up to now, the publication states that the automatic navigation of mobile robots is subject to several control techniques. The PID controller is frequently used and has a minimalistic construction, but a strong modeling approach is required. It is therefore inappropriate for nonlinear dynamic robotic devices and unsafe surroundings. Smart controllers can be regarded as the most interesting embedded systems that allow and operate the robot in the actual world. In a collection of language structures, the FL control combines the human understanding of automobile operation. Due to its simplicity and efficiency, it is extensively utilized as vehicle control for autonomous vehicles as a model that incorporates human expertise. For learning and adaptability, it can be combined with NN or GA. The most robust strategy for autonomous mobile robotic navigating was the neuro-fuzzy algorithm as part of adaptive FL techniques.

The literature cited shows a growing prevalence in the farming of the autonomous navigational system for mobile robots. More study must nevertheless be carried out to develop the technologies, eliminate the constraints of fully autonomous farm vehicles and reduce costs.

References

1. Ahamed, T., Kulmutiwat, S., Thanpattranon, P., Tuntiwut, S., Ryozo, N., Takigawa, T.: Monitoring of plant growth using laser range finder. In: 2011 ASABE Annual International Meeting, Louisville, Kentucky, Chapter Number 1111373 (2011)
2. Ampatzidis, Y., Vougioukas, S., Bochtis, D., Tsatsarelis, C.: A yield mapping system for hand-harvested fruits based on RFID and GPS location technologies: field testing. Precis. Agric. **10**, 63–72 (2009)
3. Andersen, J.C., Ravn, O., Andersen, N.A.: Autonomous rule-based robot navigation in orchards, In: 7th IFAC Symposium on Intelligent Autonomous Vehicles, Lecce, Italy, pp. 43–48 (2010)
4. Åstrand, B., Baerveldt, A.: A vision-based row-following system for agricultural field machinery. Mechatronics **15**, 251–269 (2005)
5. AuatCheein, F., Steiner, G., Perez Paina, G., Carelli, R.: Optimized EIF-SLAM algorithm for precision agriculture mapping based on stems detection. Comput. Electron. Agric. **78**, 195–207 (2011)
6. Barawid, O., Jr., Mizushima, A., Ishii, K., Noguchi, N.: Development of an autonomous navigation system using a two-dimensional laser scanner in an orchard application. Biosys. Eng. **96**, 139–149 (2007)
7. Ayala, M., Soria, C., Carelli, R.: Visual servo control of a mobile robot in agriculture environments. Mech. Based Des. Struct. Mach. **36**, 392–410 (2008)
8. Benson, E., Reid, J., Zhang, Q.: Machine vision-based guidance system for agricultural grain harvesters using cut-edge detection. Biosys. Eng. **86**, 389–398 (2003)
9. Benson, E., Stombaugh, T., Noguchi, N., Will, J., Reid, J.: An evaluation of a geomagnetic direction sensor for vehicle guidance in precision agriculture applications. In: 1998 ASAE Annual International Meeting, Orlando, FL, Chapter Number 983203 (1998)
10. Billingsley, J., Schoenfisch, M.: The successful development of a vision guidance system for agriculture. Comput. Electron. Agric. **16**, 147–163 (1997)
11. Cho, S., Lee, J.: Autonomous speedsprayer using differential global positioning system, genetic algorithm and fuzzy control. J. Agric. Eng. Res. **76**, 111–119 (2000)
12. Christiansen, M.: Localization in orchards using extended Kalman filter for sensor-fusion. Master thesis, University of Southern Denmark (2011)
13. Gao, F., Xun, Y., Wu, J., Bao, G., Tan, Y.: Navigation line detection based on robotic vision in natural vegetation-embraced environment. In: 2010 3rd International Congress on Image and Signal Processing (CISP), pp. 2596–2600 (2010)
14. Gée, C., Bossu, J., Jones, G., Truchetet, F.: Crop/weed discrimination in perspective agronomic images. Comput. Electron. Agric. **60**, 49–59 (2008)
15. Gottschalk, R., Burgos-Artizzu, X.P., Ribeiro, A., Pajares, G.: Real-time image processing for the guidance of a small agricultural field inspection vehicle. Int. J. Intell. Syst. Technol. Appl. **8**, 434–443 (2010)
16. Grift, T., Zhang, Q., Kondo, N., Ting, K.: A review of automation and robotics for the bioindustry. J. Biomechatronics Eng. **1**, 37–54 (2008)
17. Guivant, J.E., Masson, F.R., Nebot, E.M.: Simultaneous localization and map building using natural features and absolute information. Robot. Auton. Syst. **40**, 79–90 (2002)

18. Hagras, H., Callaghan, V., Colley, M.: Learning and adaptation of an intelligent mobile robot navigator operating in unstructured environment based on a novel online fuzzy-genetic system. Fuzzy Sets Syst. **141**, 107–160 (2004)
19. Hamner, B., Singh, S., Bergerman, M.: Improving orchard efficiency with autonomous utility vehicles. In: 2010 ASABE Annual International Meeting, Pittsburgh, PA, Chapter Number 1009415 (2010)
20. Debain, C., Malartre, F., Delmas, P., Chapuis, R., Humbert, T., Berducat, M.: Smart range finder and camera fusion-application to real-time dense digital elevation map estimation. In: Robotics 2010, International Workshop of Mobile Robotics for Environment/Agriculture, Clermont-Ferrand, France (2010)
21. Ding, Y., Chen, D., Wang, S.: The mature wheat cut and uncut edge detection method based on wavelet image rotation and projection. Afr. J. Agric. Res. **6**, 2609–2616 (2011)
22. Eaton, R., Katupitiya, J., Siew, K.W., Howarth, B.: Autonomous farming: modelling and control of agricultural machinery in a unified framework. Int. J. Intell. Syst. Technol. Appl. **8**, 444–457 (2010)
23. Ericson, S., Astrand, B.: Row-detection on an agricultural field using omnidirectional camera. In: 2010 IEEE/RSJ International Conference on Intelligent Robots and Systems (IROS), p. 49824987 (2010)
24. Jiang, H., Xiao, Y., Zhang, Y., Wang, X., Tai, H.: Curve path detection of unstructured roads for the outdoor robot navigation. Math. Comput. Model. (2011)
25. Joshi, M.M., Zaveri, M.A.: Reactive Navigation of autonomous mobile robot using neuro-fuzzy system. Int. J. Robot. Autom. (IJRA) **2**, 128–145 (2011)
26. Han, S., Zhang, Q., Noh, H.: Kalman filtering of DGPS positions for a parallel tracking application. Trans. Am. Soc. Agric. Eng. **45**, 553–560 (2002)
27. Han, S., Zhang, Q., Ni, B., Reid, J.: A guidance directrix approach to vision-based vehicle guidance systems. Comput. Electron. Agric. **43**, 179–195 (2004)
28. Hansen, S., Bayramoglu, E., Andersen, J.C., Ravn, O., Andersen, N., Poulsen, N.K.: Orchard navigation using derivative free Kalman filtering. In: 2011 IEEE Proceedings of the American Control Conference (ACC), pp. 4679–4684 (2011)
29. Harper, N., McKerrow, P.: Recognising plants with ultrasonic sensing for mobile robot navigation. Robot. Auton. Syst. **34**, 71–82 (2001)
30. Hellström, T.: Autonomous navigation for forest machines. A project pre-study in the Department of Computer Science Umea University, Sweden (2002)
31. Iida, M., Burks, T.: Ultrasonic sensor development for automatic steering control of orchard tractor. In: Proceedings of the Automation Technology for Off Road Equipment, Chicago, Illinois, p. 221229 (2002)
32. Kaizu, Y., Imou, K.: A dual-spectral camera system for paddy rice seedling row detection. Comput. Electron. Agric. **63**, 49–56 (2008)
33. Libby, J., Kantor, G.: Deployment of a point and line feature localization system for an outdoor agriculture vehicle. In: 2011 IEEE International Conference on Robotics and Automation (ICRA), Shanghai, China, pp. 1565–1570 (2011)
34. Kise, M., Zhang, Q., Rovira Más, F.: A stereovision-based crop row detection method for tractor-automated guidance. Biosys. Eng. **90**, 357–367 (2005)
35. Kise, M., Noguchi, N., Ishii, K., Terao, H.: The development of the autonomous tractor with steering controller applied by optimal control. In: Proceedings of the Conference on Automation Technology for Off-Road Equipment, Chicago, USA, pp. 367–373 (2002)
36. Kurashiki, K., Fukao, T., Ishiyama, K., Kamiya, T., Murakami, N.: Orchard traveling UGV using particle filter-based localization and inverse optimal control. In: Proceedings of the 2010 IEEE/SICE International Symposium on System Integration (SII). IEEE, pp. 31–36 (2010)
37. Leemans, V., Destain, M.-F.: Line cluster detection using a variant of the Hough transform for culture row localisation. Image Vis. Comput. **24**, 541–550 (2006)
38. Li, M., Imou, K., Wakabayashi, K., Yokoyama, S.: Review of research on agricultural vehicle autonomous guidance. Int. J. Agric. Biol. Eng. **2**, 116 (2009)

39. Anuradha, K.K., Singh, S.: Two stage classification model for crop disease prediction. Int. J. Comput. Sci. Mob. Comput. **4**(6), 254–259 (2015)
40. Lulio, L.C., Tronco, M.L., Porto, A.J.: Cognitive-merged statistical pattern recognition method for image processing in mobile robot navigation. In: Proceedings of the 2012 Brazilian Robotics Symposium and Latin American Robotics Symposium. IEEE, pp. 279–283 (2012)
41. Marichal, G., Acosta, L., Moreno, L., Mendez, J., Rodrigo, J., Sigut, M.: Obstacle avoidance for a mobile robot: a neuro-fuzzy approach. Fuzzy Sets Syst. **124**, 171–179 (2001)
42. Martín, D., Guinea, D., García-Alegre, M., Villanueva, E., Guinea, D.: Fuzzy steering control of a hydraulic tractor and laser perception obstacle avoidance strategies. In: Proceedings of the XV CongresoEspanolSobreTecnologıas y Logica Fuzzy (ESTYLF), Huelva, Spain, pp. 349–355 (2010)
43. Nagasaka, Y., Umeda, N., Kanetai, Y., Taniwaki, K., Sasaki, Y.: Autonomous guidance for rice transplanting using global positioning and gyroscopes. Comput. Electron. Agric. **43**, 223–234 (2004)
44. Okamoto, H., Hamada, K., Kataoka, T., Terawaki, M., Hata, S.: Automatic guidance system with crop row sensor. In: Proceedings of the Automation Technology for Off-road Equipment, Chicago, USA, pp. 307–316 (2002)
45. Noguchi, N., Kise, M., Ishii, K., Terao, H.: Field automation using robot tractor. In: Proceedings of the Automation Technology for Off-Road Equipment, Chicago, USA, pp. 239–245 (2002)
46. Nørremark, M., Griepentrog, H.W., Nielsen, J., Søgaard, H.T.: The development and assessment of the accuracy of an autonomous GPS-based system for intra-row mechanical weed control in row crops. Biosys. Eng. **101**, 396–410 (2008)
47. Ortiz, J.M., Olivares, M.: A vision-based navigation system for an agricultural field robot. In: IEEE 3rd Latin American Robotics Symposium, 2006, LARS'06, pp. 106–114 (2006)
48. Rovira-Mas, F., Han, S., Wei, J., Reid, J.: Fuzzy logic model for sensor fusion of machine vision and GPS in autonomous navigation. In: 2005 ASAE Annual International Meeting, Tampa, FL, USA, Chapter Number 051156 (2005)
49. Rovira-Más, F.: Sensor architecture and task classification for agricultural vehicles and environments. Sensors **10**, 11226–11247 (2010)
50. Rovira-Más, F., Zhang, Q., Reid, J.F.: Stereo vision three-dimensional terrain maps for precision agriculture. Comput. Electron. Agric. **60**, 133–143 (2008)

AR/VR Technology for Autonomous Vehicles and Knowledge-Based Risk Assessment

Sneh Shanu, Dev Narula, Nayana, Laxmi Kumari Pathak, and Shalini Mahato

Abstract The advancement of Artificial Intelligence (AI), which includes Augmented Reality (AR) and Virtual Reality (VR), has sped up the continuous advancement and deployment of Autonomous Vehicles (AVs) in the industry of transportation. This project is powered by big data from numerous sensing devices as well as modern computational resources. AI has become an integral component of AVs for understanding the surrounding environment and making suitable decisions while in motion. It's critical to understand how AR and VR work in AV systems in order to achieve the goal of full automation (i.e., self-driving). Existing research has put a lot of effort into looking into various aspects of using AR and VR in AV development. However, few studies have provided a thorough examination of current practices in implementing AR and VR in AVs to the research community. As a result, by providing a comprehensive survey of key studies in this research area, this chapter aims to close the gap. Its specific purpose is to estimate the danger of adopting AR/VR technology in the automotive industry since real-time risk assessment of autonomous driving at operational and tactical levels is exceedingly difficult because both contextual and circumferential aspects must be considered simultaneously.

Keywords Augmented reality (AR) · Virtual reality (VR) · Autonomous vehicles (AV) · Risk assessment · Knowledge-based risk assessment

1 Introduction

Augmented reality and Virtual reality are surging through the tech sphere in 2020, particularly in the automobile industry. Any driver (and any other pedestrian on the road) can be put in danger by a momentary glance at the dashboard or side mirror or lack of concentration on the road, AR/VR play a major role in eliminating such hazards because of the advancement of this cutting-edge technology, manufacturers

S. Shanu · D. Narula · Nayana · L. K. Pathak (✉) · S. Mahato
Amity Institute of Information Technology, Amity University Ranchi, Ranchi, Jharkhand, India
e-mail: lkpathak@rnc.amity.edu

A. E. Hassanien et al. (eds.), *Virtual and Augmented Reality for Automobile Industry: Innovation Vision and Applications*, Studies in Systems, Decision and Control 412, https://doi.org/10.1007/978-3-030-94102-4_5

are now able to mitigate these risks, ensuring a safer and more comfortable ride every time you get behind the wheel.

Augmented Reality is a technological advancement that is revolutionising the way we interact with the world. AR works by superimposing digital information such as graphics, photos, or text over our regular first-person vision. Wearable technology, such as glasses or goggles, or projection make this possible. While some wearable technology allows us to have more access to our digital lives, AR allows for complete immersion that is not possible with a phone or smartwatch by itself. This way, you can get real-time access to both your digital and surrounding information.

For firefighters whose vision is impaired by smoke, fire and low ambient light levels, a vibro-tactile glove might offer details regarding the distance of prominent items in poor vision search situations. A jaw-worn microphone-audio and a head-worn display combination can be utilised to offer real-time audio and visual augmentation, such as neutralising surrounding sounds and boosting the chosen voice, or detecting and overlaying movement to assist users perceive traces and patterns unseen to the human eye. By seamlessly interweaving multimedia information with real-world objects, these systems enable us to go beyond the capacity of normal-functioning human sensory systems [14].

In the near future, autonomous vehicles, whether in the air, on land, or at sea, plainly provide a huge commercial potential. According to Goldman Sachs Research (2015), the drone business will be worth $100 billion between 2016 and 2020. This comprises $70 billion for military uses, $17 billion for retail, and $13 billion for commercial/civil applications. Construction, agriculture, insurance claims, and public safety are among the most important commercial/civil industries (police, fire, coast guard). Driverless automobiles and drone package delivery systems are two examples of autonomous vehicle applications [15]. The uses of AR/VR in various industries are shown in Fig. 1.

Unlike Virtual Reality, however the goal of Augmented Reality is not to alter the information that the user is processing. AR allows people to merge the physical and digital worlds rather than creating a completely separate digital world.

Virtual Reality can transform how one can learn to drive a car on their own. You can pick between the manual or automatic driving experience in Virtual Reality as well. For example, take a virtual reality game that teaches you to drive; these are pretty much more or less perfect driving simulators. VR allows vehicle production manufacturers to test and evaluate their cars under any circumstances such as different climates, weather conditions and in any location. Vendors can use the immersive technology to ensure a high level of safety for their autonomous vehicles while saving money and time.

A large quantity of training data was required to train a CNN for dynamic risk assessment using supervised learning. It's difficult to generate this much data in a real-world driving situation. As a result, we employed a simulation environment to generate enough training data. To be reflective of a genuine world, this ecosystem must be vast and random. Computer games are frequently used as a replacement for big-budget professional simulation tools in research [8].

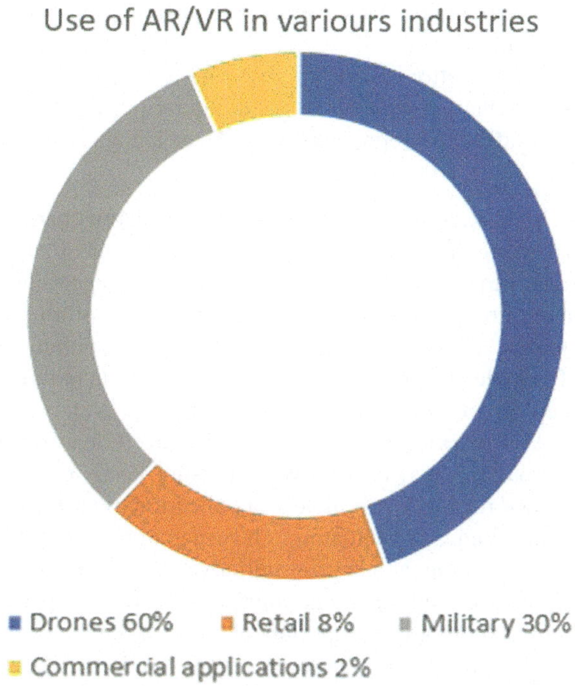

Fig. 1 Use of AR/VR in various industries

In automobiles, augmented reality is becoming an intriguing way to improve active safety while driving. A possible application is directing a driver's attention to an impending hazard anywhere around the automobile. In the present state of driving technology, it is critical that drivers have easy access to driving information. Many modern electronic gadgets, such as an autonomous navigation system and a lane departure warning system, are utilised for driving safety support [1].

Safety is a top priority for both Advanced Driver Assistance Systems (ADAS) and Autonomous Vehicles. Risk is a closely related notion that can be thought of as the probability and severity of future harm to a vehicle of interest. It is evident from this description that mathematical models that allow us to forecast how a situation will evolve in the future are required in order to estimate the risk associated with a certain circumstance. This paper looks at some of these models and how they relate to risk assessment. Motion modelling and prediction approaches are categorized based on the types of hypotheses they generate about the entities being modelled. The following are the various motion models:

i. The simplest models are physics-based motion models, which assume that vehicle motion is only determined by physical rules. They are discussed in Section 'Physics-based motion models.
ii. Manoeuvre-based motion models are more advanced since they evaluate how a vehicle's future motion is affected by the manoeuvre the driver intends to

undertake. These models are presented in Section 'Manoeuvre-based motion models.

iii. Interaction-aware motion models consider the interdependencies among vehicle manoeuvres. In the literature, there are only a few examples of such advanced representations. They are discussed in Section 'Interaction-aware motion models [1].

Risk assessment is concerned with a distinct part of the problem: how to measure risk using these motion models. Due to the fact that the actual meaning of the word "risk" differs across them, we propose categorising existing techniques into two major groups. The first considers just the possibility of physical collisions between entities. The second set of approaches incorporates the concept of risk as a result of vehicles responding differently than predicted in a particular environment (e.g., according to traffic rules).

The robotics technique and the deep learning approach are the two main approaches used to achieve increasing levels of automation. The classic robotics method regards the vehicle as a collection of various roles accomplished by sensors, control logic, and actuators. The usage of augmented reality in vehicles is becoming more common, and several initiatives are investigating this technology as a method to enhance road safety and the general driving experience. Traffic signs, also known as road signs, are an important element of the road environment because they convey visual signals to all road users, not just drivers [1].

2 Background

In the beginning of the twentieth century the computational capabilities were very limited, the tasks performed on our electronic devices now which we take for granted was once a difficult concept to imagine with the level of hardware available at the time. GUI's, graphical and 3D rendered images did not exist until latter half of the twentieth century. twentieth century saw rapid advancement in technology, modern day GUI and Graphic designing became available due to advancements in technology back then, this also gave birth to two very important technologies of the modern-day computing Augmented and Virtual reality. AR and VR are two very rapidly growing technologies that use the concept of digital environment.

The first VR/AR headset was created by Ivan Sutherland also called the "*father of computer graphics*" along with his student Bob Sprull in the year 1968 [3]. It used 2 6DOF trackers to display images [28], the technology of that time was very limited in terms of computing power and only capable of displaying very simple images such as that of a wireframe drawing. The term "*Augmented Reality*" was later coined in the year 1992 by Thomas P. Caudell and David Mizell [3]. Advancements in VR were already in progress since the mid of 1830s and major improvements were seen in the form of "*Sensorama*" *(1962)* and Ivan Sutherland's head-mounted display

[11, 28]. The term "*Virtual Reality*" was coined and popularized by Jaron Lanier in the year 1987 [33].

In the beginning AR/VR were mostly used in the fields of science and research, flight simulation and to train pilots for combat, military purposes and space exploration [11]. Earlier models were very bulky and very expensive making them out of reach for the normal consumers. In the recent years the use of AR/VR in our lives has increased drastically. VR has seen a major boom in the entertainment, education and gaming Industries. Google carboard which can be bought for around 20$ makes it easily accessible for almost anyone along with numerous other cheap DIY headsets. AR was also popularized with the launch of the hugely popular game "*Pokemon GO*" in the year July 2016 [21]. Eventually AR and VR became widespread in the automotive domain being used for a training of new employees, self-driving cars, HUD's, virtual testing etc. Few of the applications of AR/VR in autonomous vehicles are shown in Fig. 2.

i. **Navigation**—Car navigation systems are becoming more common in today's automobiles. However, information systems that are based on location are no longer tied to automobiles. Portable outdoor navigation systems for walkers and digital tourist guides are already available from Personal Digital Assistant.

   For assisting with car navigation, there are three types of Augmented Reality (AR) displays:

   a. Heads Up Display (HUD)
   b. Head Mounted Display (HMD) and

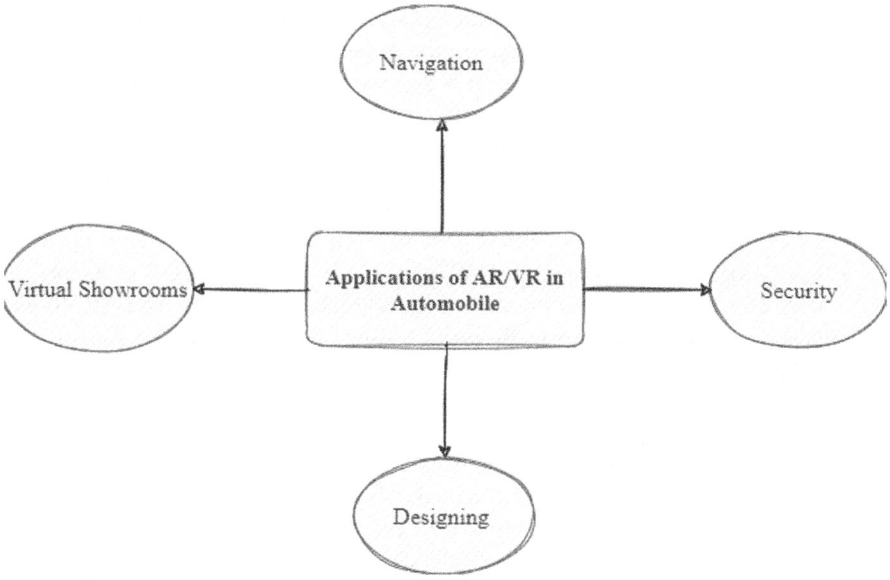

Fig. 2 Application of AR/VR in autonomous vehicles

c. Heads Down Display (HDD).

The Head Down Display (HDD) on the dashboard shows an AR-enhanced live video view. Augmented reality (AR) cues are superimposed on a live camera feed of the real world in this display. The driver's line-of-sight Heads Up Display (HUD) on the windscreen is another example. This reduces the amount of time the driver must look away from the road and gives them a larger field of vision. Google Glass and other head-mounted displays (HMDs) are used to create a personal virtual display that overlays the real world. The main benefit of wearing a head-mounted display is that it allows the user to see navigation information regardless of where they are looking [20].

With an overwhelming amount of information presented on the auto navigator's screen, the user's attention may become overly concentrated on the screen. As a result, the automobile navigation display should be built to limit the amount of information displayed, allowing for quick detection of information that is useful for route navigation [2].

ii. **Security**—Car owners nowadays are concerned about their automobiles being stolen from a parking spot, parking lot or driveways. An image/photograph processing-based real-time car theft detection and prevention system can solve this problem. There are various low-cost smart automobile security systems on the market, including FDS (Face Detection Subsystem), GPS (Global Positioning System), and GSM (Global System for Mobile Communications).

One conceivable technical option is a GSM-based car/vehicle security system, which has been created by many organisations to identify the car/vehicle position after it has been robbed. The GPS module is used to trace the vehicle's current position, which is relayed to the owner's phone through the GSM module. The face recognition entails matching the input face to the database's faces. Face recognition is done using the PCA method [25].

iii. **Designing**—Today's automobile designers have the option of building their vehicles from a variety of pre-existing components. Because there are so many components and variations, it is impossible to build a prototype for each design option. Augmented Reality overlays fresh designs on an existing structure, assisting the designer in making decisions. The prototype's design is based on the VRMl2.0 visual standard. CATIA, a CAD system, is used to export the files. The application employs a variety of tracking methods. The AR-Toolkit places a virtual vehicle in a real-world setting on a table. The user may use his hands to interact with the virtual item, such as picking up different components of the automobile such as the front door, backdoor, seats, or grasping a mirror and moving it to a new place in the car [10].

iv. **Virtual Showrooms**—If you've ever considered purchasing a new vehicle, you've almost certainly gone to a dealership to test drive it. However, you've had very little opportunity to test drive a car with the exact set of features you're looking for. Official automobile dealers generally have a limited number of car models available. Because a store's space is restricted, importing more models isn't a viable option. Virtual reality reduces the need to hire larger spaces and stock dealerships with a huge number of vehicles. Customers may customise their automobile and observe every aspect in a virtual world using a VR HMD, choosing from a variety of equipment options. Virtual reality allows car dealers to display their consumers what the future will be like, making it an ideal marketing tool.

3 Literature Survey

Annually, more than 1 million people die and 50 million people are wounded in road traffic accidents throughout the world, according to data. Only in 2019, there are 164,358 roads. In the Russian Federation, 16,981 persons were killed as a consequence of accidents. 210,877 people were killed, and another 210,877 were wounded. In this case, there were 146,688 accidents caused by traffic infractions, with 14,420 persons died. As a result, it is evident that traffic infractions account for more than three-quarters of all accidents. Based on the information supplied by the portal stat.gibdd.ru, it can be concluded that in order to maintain road safety, it is required to design and implement a set of measures to enhance traffic skills and knowledge. The expansion of the urban road network is not the only answer to the challenges mentioned. There are a variety of intelligent transportation systems available. Conditions such as traffic speed, duration of the original accident, hourly volume, rainfall intensity, and number of cars involved in the primary accident were used to evaluate the parameters related with subsequent accident chance. The method incorporates a variety of modelling techniques, including motion-based driving simulations, pedestrian simulations, motorcycle and bicycle modelling, and traffic modelling. Driving a car safely necessitates the processing of enormous quantities of dynamic data while under time constraints. Drivers, on the other hand, can only pay attention to a limited fraction of visual stimuli at a time [29].

Eco-Driving

Eco-driving is a cutting-edge form of transportation that emphasizes fuel efficiency, speed, and safety. Some of the features of eco-driving are shown in Fig. 3.

i. **Driving speed**:

   A steady pace should be maintained when driving. Maintaining a speed between 60 and 90 km/h (37–56 mph) results in a significant reduction in emissions per unit distance [7].

   There was also a minor range of 50–70 km/h (31–43 mph) identified for reduction in emission per unit distance. These are, of course, fairly wide ranges; more specific figures may be calculated based on engine type, vehicle mass, and a variety of other factors. Furthermore, because the numbers are considerably below existing highway speed restrictions, adherence to these limitations would be confined to urban roads [30].

Fig. 3 Features of Eco-driving

ii. **Acceleration and deceleration**

From a fuel-saving standpoint, it is usually recommended to avoid aggressive driving techniques in favour of smoother ones. Maintaining a safe distance from the vehicle in front of you, anticipating traffic lights and unexpected events, avoiding unnecessary accelerations and decelerations, and keeping as close to a goal speed as possible are all ways to achieve this. However, it is not always the case that modest accelerations are preferred.

iii. **Route choice**

Despite the fact that it isn't really an eco-driving advice, it has a significant impact on consumption and emissions. The shortest or fastest route is suggested by navigation systems on the market, but none of them take into account the amount of gasoline required to get to the destination. In 46% of situations, proposed routes were not the most ecologically friendly, and a fuel-optimized navigator might save 8.2% of the time.

iv. **Weight**

Extra loads should be kept to a bare minimum, given that 45 kg might increase fuel consumption by 1–2% on a small car. Driving with under-inflated tyres causes a comparable rise.

v. **Idling**

It ought to be avoided or reduced in a zero-fuel efficiency condition, since it is defined as the act of remaining motionless in neutral gear. Furthermore, when a vehicle is idling, a greater number of pollutants, such as carbon oxides (CO), hydrocarbons (HC), nitrogen oxides (NOx), and particulate matter (PM), are generated (PM). Many cars have included the "start and stop" capability in the last few years to automate the switching on and off of the engine, with good results, especially while travelling in an urban area. For cars that do not start and stop, it is predicted that this technology may save up to 20% CO_2. It's also enabled by the fact that contemporary motors do not have multiple ignitions, thus idle durations longer than 10 s may be replaced with a switched off phase [9].

vi. **Air conditioning**

The compressor of an air conditioner may require up to 5–6 kW of electricity. When compared to medium-size automobiles, it's roughly the equivalent of driving at 55 km/h. This implies that because the aerodynamic drag isn't too severe, it's easy to roll down the windows at moderate speeds. Instead, the drag rises, causing the air conditioner to switch on at greater speeds, resulting in lower usage.

S. No	References	Literature review	Conclusion/Result
1	Bran et al. [5]	The 3D AR visualisation technique was proposed by the authors that improves on the in-vehicle system's automated adaptive notification filtering module. Authors used voice and gesture commands to manage standard driving tasks like navigation system operations, radio or air conditioner management, and generic activities like responding to incoming alerts. The technology is put to the test in a lab setting, with a simulated prototype car deck and windshield	Authors described about the conceptual design considerations that went into the creation of the AR-based visualisation component, which is intended to complement the adaptive filtration of smartphone-related alerts module that was previously presented in
2	Ma et al. [17]	The authors of this publication try to bridge the gap by giving a comprehensive overview of significant works in this field. Its goal is to look at how AIs are used to assist the three main functions in AVs: perception, positioning, and mapping, and strategic planning. This article also offers insights on possible prospects for using AI in conjunction with other developing technologies, based on an examination of existing practises and technological advancements: (1) high-definition maps, amount data, and high-performance computation; (2) improved simulation platform AR/VR; and (3) 5G connectivity for linked AVs	In the context of developing technologies, future activities that can assist supplement the usage of AI for aiding AV development have been identified: (1) big data, HPC, and a high-resolution digital map for better data gathering and processing; (2) AR/VR enhanced platform for building accelerated test scenarios; and (3) 5G for low-latency connectivity between AVs

(continued)

(continued)

S. No	References	Literature review	Conclusion/Result
3	Shi et al. [26]	This paper provides a domain-specific automated machine learning (AutoML) model for predicting risk and assessing behaviour that may be utilised in autonomous vehicle behavioural decision-making and movement trajectory planning (AVs). The AutoML integrates three major aspects: unsupervised risk identification by surrogate key risks and big data clustering, feature learning based on XGBoost, and model auto-tuning by Bayesian optimisation, which enables end-to-end machine learning from speed control and sensing data to detailed risk levels and corresponding different senses. The functionalities and performance of AutoML are then assessed using NGSIM data and various sensor setups or data collecting situations as assumptions	AutoML delivers good results in behavior-based risk prediction, with an overall predictive power of 91.7% and a saferisk distinction accuracy of more than 95%. The predictive capabilities of various AV sensor setups were assessed using a unified AutoML process, which provides data-driven insights into AV safety from the perspective of the necessary information for risk-based decision-making. Furthermore, the AutoML identifies the most essential behaviour characteristics for risk assessment, revealing vital information regarding sensor data mining. Unsupervised hazard identification using FCM clustering and associated risk, threat feature learning using a hybrid of XGBoost-based filtering and RFE, highly imbalanced down sampling, model selection, and hyper-parameter tuning using Bayesian optimisation are just a few of the features that the AutoML integrates into an auto-optimisable pipeline

(continued)

(continued)

S. No	References	Literature review	Conclusion/Result
4	Wang et al. [31]	The authors of this study presented a new Rear-end Accident Prediction Mechanism using Deep Learning Method (RCPM), which uses a model of a convolutional neural network. To address the problem of class imbalance, the dataset is smoothed and extended using evolutionary theory in RCPM. The convolutional neural network model used by the authors was trained using the precompiled dataset, which was split into training and testing sets. The results demonstrate that, when compared to Honda, Berkeley, and multi-layer perceptive neural-network-based methods, RCPM significantly increases performance in predicting rear-end crashes	A novel prediction method based on convolutional neural networks to tackle the problem of Back-end collision in real time warning. Authors developed a 1.5-s learning algorithm that takes real-time traffic data and predicts it. Authors split the real vehicle trajectory NGSIM into a training set and a testing set to train and evaluate the technique in this study. They Smoothed and expanded the data, as well as converting each 1.5 s of text data into a three-dimensional pixel matrix, solving the problem of class imbalance in the training set. Authors enhanced the prediction model's performance iteratively
5	Bautu et al. [4]	The authors present an in-vehicle software solution that makes it easier for the driver to communicate with his or her smartphone. While driving, the driver can use the proposed software system to automatically filter messages and notifications from third-party applications in an appropriate manner. The technology detects potentially harmful circumstances using data from the vehicle's sensors and adjusts its behaviour to minimise human attention. The separation of concerns principle governs the software development process. The system adheres to Intelligent Environments principles, particularly in terms of putting the user's safety first at all times	They introduced IV-SFDA in this article, a software system that allows drivers to handle in-vehicle smartphone-related details without having to use a mobile device by hand. The system follows certain important Intelligent Environments concepts, such as honouring the requirements and preferences of individuals it is assisting in achieving their goals without requiring specific technical expertise from the user. Authors emphasise that driver safety is the primary issue for which their system was designed; as a result, principle (P6) is also adhered to

(continued)

(continued)

S. No	References	Literature review	Conclusion/Result
6	Calvi et al. [6]	The possibility of employing Augmented Reality (AR) technologies to boost pedestrian crossing safety was investigated in this research, using virtual information supplied to a motorist approaching a zebra crossing location. Driving simulator research was conducted to attain this goal. The system's efficacy was evaluated, and the results of simulation testing with and without AR warnings to alert the driver of a crosswalk ahead were analysed. AR alerts were evaluated in two distinct scenarios: a visible pedestrian crossing the road and a non-visible individual preparing to cross the road, which was rendered invisible by specific barriers. This study revealed the significant benefits that augmented reality and linked car technology may have on overall road safety, particularly in high-risk circumstances and complex manoeuvres. The driving simulator is an excellent tool for researching and assessing such technologies, as well as their effects on driving performance	By putting 46 volunteers through a series of tests in the LASSTRE driving simulator, this research demonstrated the potential of Augmented Reality technology to increase pedestrian crossing safety Two different events were tested to determine the efficacy of a proposed AR technology consisting of a flashing red arrow over a pedestrian and the same visual prudence but with an additional audible warning: a visible pedestrian crossing the road and a pedestrian obstructed by certain obstacles. The impacts of AR alerts were investigated by comparing vehicle performance (decelerations, speeds, and distances) as well as the substitute safety measures TTC (Time-to-Collision) and TTZ (Time-to-Zebra) with those recorded in the baseline period (without an AR warning). To prevent skewing the results owing to driver attitudes, driving dynamics, age, stress level, emotional state, neuro-cognitive state, or other factors, a homogenous sample of participants was chosen and the same standard simulation methodology was used

(continued)

(continued)

S. No	References	Literature review	Conclusion/Result
7	Hagele and Sarkheyli-Hagele [12]	To aid decision-making and action selection, this article tackles the topic of hazard detection and risk assessment in open and non-predictive situations. As operational limitations, reducing situational hazards and maintaining safety are included into decision-making and action selection. In most cases, current application-related safety requirements, as well as scenario modelling and knowledge representation, are not taken into account. Authors proposed a new method called Safety-Driven Behaviour patterns Management, which focuses on scenario modelling and the challenge of knowledge representation in its many sub-functions in the context of situational hazards	In the context of autonomous driving, the study described in this paper focuses on situational danger detection and risk assessment. The current state of the art is briefly described in this contribution. A new Safety-Driven Behaviour Management system is described in full, with modules for scenario hazard detection and risk assessment Using an urban driving situation, the illustrative scenario is utilised to illustrate the functioning of this innovative technique. The discussion and analysis demonstrate the innovative approach's practicality and believability
8	Pietra et al. [22]	The design and preliminary testing of a virtual reality modern vehicle capable of delivering haptic and visual cues to improve environmentally sustainable driving behaviour are described in this article. The driving simulator was created with the Unity game engine, and a large street scene was built to test various driving behaviours, including high-speed and urban segments. The equipment included a virtual driving seat with a steering wheel and brakes, as well as an Oculus Rift headset for a fully immersive experience. The driver received haptic input (vibrations) through the accelerator pedal, while visual information (icons and colours) was displayed on a virtual head-up display. The sensory feedbacks were given separately and together, providing information concerning excessive load torque	The design approach, development phase, and early user testing of a prototype created to examine the influence of visual and haptic input on eco-driving behaviour were all described in this paper. To provide the testers with a realistic and emotionally engrossing experience, immersive VR through an HMD has been coupled with a setup consisting of a driving seat, steering wheel, and pedals. To provide a complete picture of the user-machine interaction, authors gathered driving performance measurements, usability surveys, and physical reactions

(continued)

(continued)

S. No	References	Literature review	Conclusion/Result
9	Yang et al. [32]	In this work, authors presented a lane change judgment scheme based on "Driving Environment Assessment and Decision Making (DREAM) index" utilising just commercial automobile radar sensors. The notion of a dynamic occupancy grid zone is used to measure the index associated with a risk scenario. The local area's predetermined distance was established to meet the international requirement for a steering function. In addition, the risk assessment was carried out utilising predictive modelling of relative motion with a driving environment. Second, an index connected with the surrounding cars' cooperative driving idea is proposed. They developed a discrete-time state estimator to estimate relative acceleration that is not directly detected by radar	The authors developed a simple and efficient decision approach for ALCSs based on DOZE that solely uses radar sensors in this work. CRI and CII were combined to create the DREAM index. A constant acceleration model was used to build the state estimator, and the relative motion of STVs was predicted until the ALCS was finished. They ran a real-world test with the test cars to verify the algorithm. Several testing revealed that the car using the suggested algorithm made the correct judgement and successfully changed lanes. To evaluate the decision system, they conducted scenario-based tests using test cars on a high-speed circuit. They discovered that the DREAM index could make effective judgments in the studies, and that the lane change manoeuvrings were successful in real-world tests

4 AR/VR in Autonomous Vehicles

Augmented Reality (AR) can be embedded into cars by the use of motion capturing cameras [1]. AR combines real-world and virtual content in real-time by overlaying graphical items onto or combining with the real world. The actual camera record and registers the real-world environment and adds collected virtual information to it [27]. The major goal is to provide information that will enhance and improve the user's vision of the world and change the focal point of interaction in an unobtrusive manner. Using Heads Up Displays (HUD) information can be displayed on the windscreen of a car and assist drivers [18]. AR is also a tool for road safety while driving, using AR we can detect oncoming traffic and other obstacles in the path of the vehicle and display them to the HUD to bring user attention on to them, the camera can also have features such as infrared detection and night-vision camera to improve the visualization [23], such technology can be of major benefit especially during bad weather and/or limited visibility to identify and locate other obstacles in the path and even prevent collisions from occurring by the use of AI. AI can also detect and understand road signs and warn the user. We can identify oncoming cars and their trajectory of motion and make any necessary changes in our own manoeuvring using AI decision making to steer away and prevent any damage to life/property [34]. When a possible danger is recognized, the AR system warns the driver by displaying the time to impact for the lead car as well as approaching vehicles from side lanes on an AR display which overlaps with the lead or merging vehicles [24].

In such motion modelling and predictive algorithms used to identify the vehicle pathing Vehicles are represented as independent manoeuvring entities in manoeuvre-based motion models, which presume that a vehicle's motion on the road network is the result of a series of manoeuvres performed independently of other vehicles [16] The future path of the vehicle is calculated and analysed for the reliability of model. Factors such as speed of the vehicle and oncoming cars is also taken into consideration and precise time to collision can be calculated using physics-based models as well as differentiate from close passing to unavoidable collisions. The process of detection of pedestrian is shown in Fig. 4.

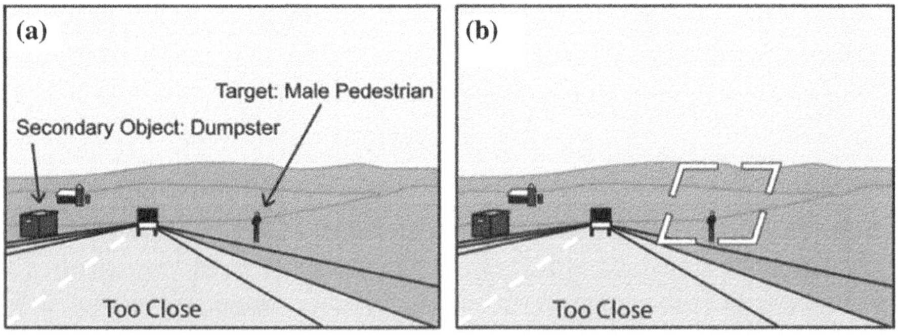

Fig. 4 Demonstrating detection of pedestrian

5 Risk Assessment of AR/VR in Autonomous Vehicles

Risk measurements are widely used in the automobile industry to create situation awareness for advanced safety features. The ISO standard for forward warning systems uses thresholds of the metrics Time to Collision (TTC) along with Enhanced Time to Collision (ETTC) to determine whether and which collision warning method should be triggered (ETTC).

In traffic conditions, a Risk Metric Calculator (RMC) is used to better understand and evaluate the values of various risk metrics. The RMC accepts driving cases that are either developed in the tool or delivered as a log from the simulation environment, fills a grid map with essential information for each simulation step, and then uses that grid pattern to calculate the selected risk metrics. In the robotics sector, grid maps are a typical notation. They divide the surroundings into fixed or variable-sized cells. The occupancy status of each cell is stored in each cell. It may contain information about an object that is holding the cell if it is occupied. This map reflects the robot's World Model, such as the vehicles, and is used to identify suitable behavior, such as the desired trajectory [8].

Autonomous Vehicles (AVs) navigate dense urban traffic, construction zones, roadway settings, residential streets where children appear and disappear by filtering through parked vehicles, segments with unpredictable traffic dynamics and traffic co-participants, and roads with traffic incidents such as vehicle traffic bottlenecks breakdowns, network deficiencies and collision hotspots by filtering through parked vehicles. Even if AVs perform all they're intended to, the fundamental safety difficulty would be figuring out how to account for these aspects in AV collision risk assessments. Currently, a motion model is used to estimate the risk of a collision by predicting the planned trajectories of other vehicles and objects in a given traffic environment and comparing them to the course of the interested AV [13].

The manner in which AR images are displayed has a considerable impact on their ability to be utilized successfully in an automobile system. The level of choice afforded to AR HUD interface designers when delivering information to drivers is still largely restricted by HUD technological limitations. The best location for AR cues in relation to the driver's forward vision is a point of contention in terms of interface design. On the one hand, virtual images centered in the driver's field of vision are likely to obscure important real-world information or impair a driver's ability to recognize danger due to the overwhelming presence of virtual images [18].

When compared to a situation where only audio is exchanged, augmented reality-based video chats between automobile drivers and distant callers generate significant visual distraction. As it turns out, augmented reality has some promise in this area, which might be a gift in disguise. However, there are several drawbacks that may turn it into a curse [19].

Despite the apparent advantages of autonomous cars, their research, particularly testing, is still in its early stages. Traditional software strategies are frequently ineffective since it is difficult to establish why the software made a particular conclusion. As a result, a greater degree of black box testing is required. However, without the full system and the capacity to sense the surroundings, it may be impossible to test sub-components of the system. This opens up the possibility of using system and environment predictions to drive black box testing, delivering stimuli to the subcomponents so that their behavior may be observed.

A simulation-based method is described that relies on VR and AR to provide external stimuli in the absence of available actual stimuli early in the process in order to encourage hardware/software codesign and development. This is accomplished by using a parallel development approach for a physical vehicle simulation and a virtual environment in which the vehicle could operate. Based on the basic needs for the physical vehicle, a behavioral model is built and simulated, allowing for initial testing of the autonomous software. The details of the vehicle are added to the model as it is created, enhancing the realism of the simulation until it is fully functioning. The procedure begins with a completely simulated virtual reality system. While the hardware system is completely simulated, the autonomous software is being evaluated in its present state, not a simulated version. According to the autonomous software development lifecycle, the present state can move from behavioral to algorithmic to functional. This is made possible by the software architecture, which isolates the autonomous software from direct interaction with the vehicle hardware, allowing information to be changed or supplemented as it passes from the hardware to the software [15].

The virtual environment is used to provide the required input data for modelling sensors as input to the sense stage or modelling sensed data provided to the world representation. This allows the system to operate in full virtual reality or to supplement reality to varying degrees with virtual information for augmented virtuality and augmented reality. However, if the virtual environment's realism is insufficient, the models employed to translate the information to the format required by autonomous software will not be able to detect a difference between virtuality and reality.

Everything begins virtual in a virtual reality (VR) setting, including the surroundings, autonomous car hardware, and software. In this setup, the autonomous software is tested during its whole lifespan. The software runs on the host computer platform at first, but it can be moved to an emulated platform for computing afterwards. Running real autonomous technology on the autonomous vehicle's physical computing platform, which necessitates the establishment of a transmission link between both the VR testbed and the autonomous vehicle, is the highest phase of testing, or final stage of testing, that can be accomplished solely through virtual reality.

Depending on the type, size, and mobility range of the autonomous system, the software architecture handles communication between the VR testbed and the autonomous system, which can be wired or wireless (Wi-Fi, Bluetooth). Because time does not exist in real autonomous vehicle operations, the VR testbed and the autonomous system must be time synced. For example, the VR testbed's dynamic objects should be updated in real time using a wall clock, and the VR testbed should be paused during communication with the autonomous vehicle.

The testing is now done in a tangible, real-world setting. The actual autonomous system, on the other hand, is put to the test in a real-world setting with virtual objects. The sensor data is synthesized using a mix of real-world and virtual items.

A virtual environment in which virtual items and autonomous agents are set in a realistic approximation of the reality and surroundings. The simulated environment gets real-time updates on the coordinates of the autonomous system and creates sensor information accordingly. In order to address the issues raised in the previous section, a communication link between the simulated/virtual world and the autonomous system must be established. The sensor data synthesis process must be fast enough to meet real-time requirements, i.e., similar to the sampling rate of real sensors, because the autonomous system operates in a real-world environment. The behavior of dynamic objects in the real world must be predicted in the virtual environment. This type of testing is best suited to very basic real-world scenarios, such as an empty room. The autonomous car is less vulnerable in the real world because the environment is so basic [15].

According to prior expert testing, the highpoint of the car ahead should be thin in MD to avoid interfering with the driving incident, and thicker in AD to be quickly detected when the driver concisely interrupts his non-driving-related exercise to check that the system has a correct understanding of the traffic situation, as well as to increase the difference between the two.

5.1 Knowledge-Based Risk Assessment

The modelling of risk patterns, the use of an inference engine for pattern recognition, and gaze control as an implementation are described in knowledge-based risk assessment.

i. **Risk Patterns**: A risk pattern is a situation's abstract description in which specific items pose a threat to the driver's vehicle or another participant in the traffic. Patterns are associated with qualitative description of the data repository. The characterization of risk variables and related risk values extends the pattern description provided in by risk patterns. The different fundamental predicates may be used to create complex patterns. The objects "kid" and "unknown item" would be assigned risk ratings in the patterns "a child approaches the lane where the automobile is travelling" and "an unknown thing is approaching quickly at a near or medium distance."
ii. **Pattern Matching**: If specific patterns are discovered, the pattern matching modules monitors them and alerts the risk evaluation. In each risk assessment cycle, an inference engine evaluates all risk patterns and generates all valid assignments. An assignment assigns constants (i.e., objects) to variables. The inference engine simply provides correct assignments in regard to the given pattern and its temporal interrelations.
iii. **Gaze Control**: Dickmann's advanced vision technologies allow intelligent cars to regulate their gaze and pay attention to specific areas of the surroundings. The method provided here detects potentially harmful things that the vision system should focus on. It can be chosen in what sequence and to what degree

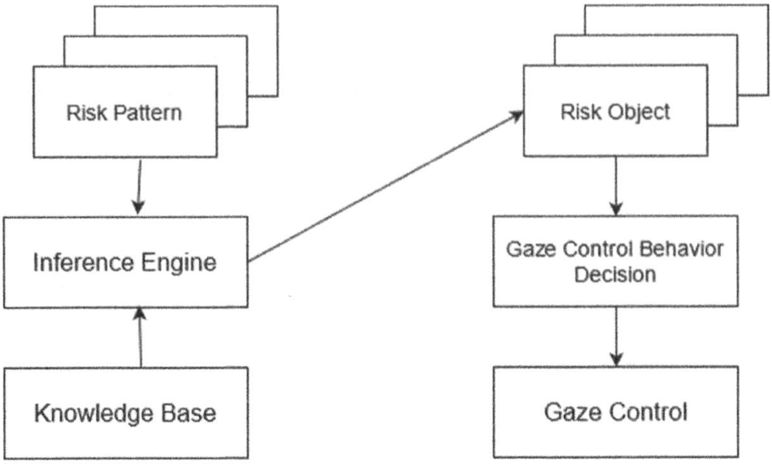

Fig. 5 Knowledge-based assessment of risk and gaze control

items should be investigated based on the risk values. The knowledge-based assessment of risk and gaze control is shown in Fig. 5.

5.2 Risk at the Vehicle Level

Describes a single car's safety context in a vehicle system (i.e., if the vehicle has the potential to collide with other vehicles). This layer's variable is likewise 'discrete,' but it takes four values to describe the safety context of every vehicle:

i. Safe driving on a section of road having safe traffic conditions
ii. Safe driving on a section of road having collision-prone traffic conditions
iii. Dangerous driving on a section of road having safe traffic conditions
iv. Dangerous driving on a section of road having collision-prone conditions.

Using these values, we can describe whether a vehicle is a potential threat to other vehicles or not. The term "Safe" denotes that the vehicle in question does not pose threat(s) to other vehicles from its manoeuvres whereas "dangerous" denotes that the manoeuvres of the vehicle in question is of potential threat to others [13].

5.3 Risk Avoidance

Avoiding risks of crash boil down to assessing the risks and then choosing methods to eliminate them for our advantage the simplest methods such as stopping the car automatically or notifying the driver ahead of time that a collision is possible and attracting their attention to the obstacle (such as a pedestrian, another vehicle or animals) can prevent most of the possible collisions if the driver interrupts, if the collision is impossible to avoid security measures to protect the driver as well as any nearby pedestrian can be initiated immediately to reduce the damage. Communication to other vehicles in the lane over a centralized network can prevent even the smallest mistake from turning in to an accident. Route geometry analysing can help identify and steer away from potholes and such and identify road marks such as pedestrian crosswalk. HUD's can also be linked to the seat of drivers to provide vibrational feedback if the driver is not paying attention on the road.

6 Discussion

This study aided in the development of a new knowledge of application of AR/VR, history and backgrounds of AR/VR technologies and AR/VR risk assessment in autonomous vehicles.

Multiple terms, including HUD, HMD, HDD, GPS, GSM FDS, bird's-eye perspective, first-person perspective, third-person perspective and bystander perspective is discussed for both VR and AR configurations to obtain a deeper grasp of autonomous systems' behavior AR/VR technologies may occasionally cause issues so risk analysis becomes necessary. A knowledge-based method for risk assessment enables the implementation of abstract rules and background knowledge, as well as the use of an inference engine to assess a situation. AR/VR is much more helpful, in comparison with the physical overview and works.

7 Conclusion

The study covers different aspects of AR/VR technology for autonomous vehicles, as well as the benefits and issues that may arise. Thus, the study would aid the researcher in the development of a new knowledge for application of AR/VR, by providing detailed history and backgrounds of AR/VR technologies and as well as present state of art-based AR/VR risk assessment in autonomous vehicles.

References

1. Abdi, L., Meddeb, A., Abdallah, F.B.: Augmented reality based traffic sign recognition for improved driving safety. In: Lecture Notes in Computer Science (Including Subseries Lecture Notes in Artificial Intelligence and Lecture Notes in Bioinformatics), vol. 9066, pp. 94–102 (2015). https://doi.org/10.1007/978-3-319-17765-6_9
2. Akaho, K., Nakagawa, T., Yamaguchi, Y., Kawai, K., Kato, H., Nishida, S.: Route guidance by a car navigation system based on augmented reality. Electr. Eng. Jpn. (English Translation of Denki Gakkai Ronbunshi) **180**(2), 43–54 (2012). https://doi.org/10.1002/eej.22278

3. Arth, C., Grasset, R., Gruber, L., Langlotz, T., Mulloni, A., Wagner, D.: The History of Mobile Augmented Reality (2015). http://arxiv.org/abs/1505.01319
4. Bautu, E., Puchianu, C.M., Bran, E., Sburlan, D.F., Popovici, D.M.: In-vehicle software system for fostering driver's attentiveness. In: 2020 15th International Conference on Development and Application Systems, DAS 2020—Proceedings, vol. 21, pp. 151–156 (2020). https://doi.org/10.1109/DAS49615.2020.9108946
5. Bran, E., Sburlan, D.F., Popovici, D.M., Puchianu, C.M., Bautu, E.: In-vehicle visualization of data by means of augmented reality. Procedia Comput. Sci. **176**, 1487–1496 (2020). https://doi.org/10.1016/j.procs.2020.09.159
6. Calvi, A., D'Amico, F., Ferrante, C., Bianchini Ciampoli, L.: Effectiveness of augmented reality warnings on driving behaviour whilst approaching pedestrian crossings: A driving simulator study. Accid. Anal. Prevent. **147**(September), 105760 (2020). https://doi.org/10.1016/j.aap.2020.105760
7. El-Shawarby, I., Ahn, K., Rakha, H.: Comparative field evaluation of vehicle cruise speed and acceleration level impacts on hot stabilized emissions. Transp. Res. Part D Trans. Environ. **10**(1), 13–30 (2005)
8. Feth, P., Akram, M.N., Schuster, R., Wasenmüller, O.: Dynamic risk assessment for vehicles of higher automation levels by deep learning. In: Lecture Notes in Computer Science (Including Subseries Lecture Notes in Artificial Intelligence and Lecture Notes in Bioinformatics), vol. 11094 LNCS, pp. 535–547 (2018). https://doi.org/10.1007/978-3-319-99229-7_48
9. Fonseca, N., Casanova, J., Valdes, M.: Influence of the stop/start system on CO_2 emissions of a diesel vehicle in urban traffic. Transp. Res. Part D Transp. Environ. **16**(2), 194–200 (2011)
10. Fruend, J., Matysczok, C., Radkowski, R. (2002). AR-based product design in automobile industry. In: ART 2002—1st IEEE International Augmented Reality Toolkit Workshop, Proceedings, p. 2. https://doi.org/10.1109/ART.2002.1106981
11. Gigante, M.A.: Virtual reality: definitions, history and applications. In: Virtual Reality Systems. ACADEMIC PRESS LIMITED (1993). https://doi.org/10.1016/b978-0-12-227748-1.50009-3. https://jasoren.com/4-virtual-reality-applications-in-the-automotive-industry/
12. Hagele, G., Sarkheyli-Hagele, A.: Situational risk assessment within safety-driven behavior management in the context of UAS. In: 2020 International Conference on Unmanned Aircraft Systems, ICUAS 2020, pp. 1407–1415 (2020). https://doi.org/10.1109/ICUAS48674.2020.9214072
13. Katrakazas, C., Quddus, M., Chen, W.H.: A new integrated collision risk assessment methodology for autonomous vehicles. Accid. Anal. Prev. **127**(January), 61–79 (2019). https://doi.org/10.1016/j.aap.2019.01.029
14. Kim, S.J., Dey, A.K.: Augmenting human senses to improve the user experience in cars: applying augmented reality and haptics approaches to reduce cognitive distances. Multimed. Tools Appl. **75**(16), 9587–9607 (2016). https://doi.org/10.1007/s11042-015-2712-4
15. Leathrum, J.F., Shen, Y., Mielke, R.R., Gonda, N.: Integrating virtual and augmented reality based testing into the development of autonomous vehicles. 31 MODSIM World (31), 1–11 (2018)
16. Lefèvre, S., Vasquez, D., Laugier, C.: A survey on motion prediction and risk assessment for intelligent vehicles. ROBOMECH J. **1**(1) (2014). https://doi.org/10.1186/s40648-014-0001-z
17. Ma, Y., Wang, Z., Yang, H., Yang, L.: Artificial intelligence applications in the development of autonomous vehicles: a survey. IEEE/CAA J. Automatica Sinica **7**(2), 315–329 (2020). https://doi.org/10.1109/JAS.2020.1003021
18. Merenda, C., Kim, H., Tanous, K., Gabbard, J.L., Feichtl, B., Misu, T., Suga, C.: Augmented reality interface design approaches for goal-directed and stimulus-driven driving tasks. IEEE Trans. Visual Comput. Graphics **24**(11), 2875–2885 (2018). https://doi.org/10.1109/TVCG.2018.2868531
19. Moghimi, M., Stone, R., Rotshtein, P., Cooke, N.: The sense of embodiment in virtual reality. Presence Teleoper. Virt. Environ. **25**(2), 81–107 (2016). https://doi.org/10.1162/PRES
20. Narzt, W., Pomberger, G., Ferscha, A., Kolb, D., Müller, R., Wieghardt, J., Hörtner, H., Lindinger, C.: Pervasive information acquisition for mobile AR-navigation systems. In:

Proceedings—5th IEEE Workshop on Mobile Computing Systems and Applications, WMCSA 2003 (Wmcsa), pp. 13–20 (2003). https://doi.org/10.1109/MCSA.2003.1240763
21. Paavilainen, J., Korhonen, H., Alha, K., Stenros, J., Koskinen, E., Mäyrä, F.: The Pokémon go experience: a location-based augmented reality mobile game goes mainstream. In: Conference on Human Factors in Computing Systems—Proceedings, 2017-May (Figure 1), 2493–2498 (2017). https://doi.org/10.1145/3025453.3025871
22. Pietra, A., Vazquez Rull, M., Etzi, R., Gallace, A., Scurati, G. W., Ferrise, F., Bordegoni, M.: Promoting eco-driving behavior through multisensory stimulation: a preliminary study on the use of visual and haptic feedback in a virtual reality driving simulator. Virtual Real. (2021). https://doi.org/10.1007/s10055-021-00499-1
23. Rizov, T., KJosevski, M., Tashevski, R.: Driver assistance systems in vehicles using augmented reality–benefits and challenges. Trans. Motauto. World **2**(5), 201–206 (2017)
24. Rusch, M.L., Schall, M.C., Gavin, P., Lee, J.D., Dawson, J.D., Vecera, S., Rizzo, M.: Directing driver attention with augmented reality cues. Transport. Res. F Traffic Psychol. Behav. **16**, 127–137 (2013). https://doi.org/10.1016/j.trf.2012.08.007
25. Shafeeq, M.: Raspberry Pi based intelligent car anti-theft system through face recognition using GSM and GPS (July) (2018)
26. Shi, X., Wong, Y.D., Chai, C., Li, M.Z.-F. (2020). An automated machine learning (AutoML) method of risk prediction for decision-making of autonomous vehicles. IEEE Trans. Intell. Transport. Syst. 1–10. https://doi.org/10.1109/tits.2020.3002419
27. Streibel, J.S., Silveira, Jr., L.G. da, Jung, C.R.: Automatic camera calibration for driver assistance systems. In: 2009 XI Symposium on Virtual and Augmented Reality (June 2017), pp. 29–36 (2009)
28. Sutherland, I.E.: A head-mounted three dimensional display. In: AFIPS '68 (Fall, part I): Proceedings of the December 9–11, 1968, Fall Joint Computer Conference, pp. 757–764 (1968). https://doi.org/10.1145/1476589.1476686
29. Tomchinskaya, T., Shaposhnikova, M., Dudakov, N.: training beginners and experienced drivers using mobile-based virtual and augmented reality. CEUR Workshop Proc. **2744**, 1–12 (2020). https://doi.org/10.51130/graphicon-2020-2-3-69
30. Wang, H., Fu, L., Zhou, Y., Li, H.: Modelling of the fuel consumption for passenger cars regarding driving characteristics. Transp. Res. Part D Transp. Environ. **13**(7), 479–482 (2008)
31. Wang, X., Liu, J., Qiu, T., Mu, C., Chen, C., Zhou, P.: A real-time collision prediction mechanism with deep learning for intelligent transportation system. IEEE Trans. Veh. Technol. **69**(9), 9497–9508 (2020). https://doi.org/10.1109/TVT.2020.3003933
32. Yang, J.H., Choi, W.Y., Chung, C.C.: Driving environment assessment and decision making for cooperative lane change system of autonomous vehicles. Asian J. Control **23**(3), 1135–1145 (2021). https://doi.org/10.1002/asjc.2455
33. Zheng, J.M., Chan, K.W., Gibson, I.: Virtual reality. IEEE Potent. **17**(2), 20–23 (1998). https://doi.org/10.1109/45.666641
34. Zhou, P., Zhang, W., Braud, T., Hui, P., Kangasharju, J.: Enhanced augmented reality applications in vehicle-to-edge networks. In: Proceedings of the 2019 22nd Conference on Innovation in Clouds, Internet and Networks and Workshops, ICIN 2019, pp. 167–174 (2019). https://doi.org/10.1109/ICIN.2019.868587

Optimal Stacked Sparse Autoencoder Based Traffic Flow Prediction in Intelligent Transportation Systems

S. Neelakandan, M. Prakash, Sanjay Bhargava, Kumar Mohan, Nismon Rio Robert, and Sachin Upadhye

Abstract Recently, intelligent transportations system (ITS) has gained significant internet due to the higher needs for road safety and competence in interconnected road network. As a vital portion of the ITS, traffic flow prediction (TFP) is offer support in several dimensions like routing, traffic congestion, and so on. To accomplish effective TFP outcomes, several predictive approaches have been devised namely statistics, machine learning (ML), and deep learning (DL). This study designs an optimal stacked sparse autoencoder based traffic flow prediction (OSSAE-TFP) model for ITS. The goal of the OSSAE-TFP technique is to determine the level of traffic flow in ITS. In addition, the presented OSSAE-TFP technique involves the traffic and weather data for TFP. Moreover, the SSAE based prediction model is designed for forecasting the traffic flow and the optimal hyperparameters of the SSAE model can be adjusted by the use of water wave optimization (WWO) technique. To showcase the enhanced predictive outcome of the OSSAE-TFP technique, a wide range of simulations was carried out on benchmark datasets and the results portrayed the supremacy of the OSSAE-TFP technique over the recent state of art methods.

S. Neelakandan (✉)
Department of Computer Science and Engineering, R.M.K. Engineering College, Chennai, India

M. Prakash
Data Science and Analytics Center, Karpagam College of Engineering, Coimbatore, India

S. Bhargava
Department of Computer Science and Engineering, Poornima College of Engineering, Jaipur, Rajasthan, India

K. Mohan
Lecturer Information Technology, University of Technology and Applied Sciences, Shinas, Oman
e-mail: kumar.mohan@shct.edu.om

N. R. Robert
Department of Computer Science, Christ University, Bangalore 560029, India
e-mail: nismon.rio@christuniversity.in

S. Upadhye
Department of Computer Application, Shri Ramdeobaba College of Engineering and Management, Nagpur, Maharashtra, India
e-mail: upadhyesd@rknec.edu

A. E. Hassanien et al. (eds.), *Virtual and Augmented Reality for Automobile Industry: Innovation Vision and Applications*, Studies in Systems, Decision and Control 412, https://doi.org/10.1007/978-3-030-94102-4_6

Keywords Traffic flow prediction · Machine learning · Intelligent transportation system · Urban traffic flow · Parameter optimization

1 Introduction

In current world, effective and smooth logistics and transportation of individuals play an essential role in guaranteeing the standard operation of the social economy. But, despite the increasing demands for transportation resources from social evolution, the current transportation structure, depending on the traditional traffic control systems (for example, the well-known loop detector-based SCOOT technique), doesn't work efficiently [1]. Traffic congestion plays an important role in our day-to-day life. Also in 2011, the consistent loss further improved to over 25 million tons of automotive exhaust emission, $121 billion costs in fuel loss, and 5.5-billion-hour delay [2]. Thus, in order to efficiently use the current transport structure to mitigate the conflicts among the demands for the movement of goods and transportation resources, has gradually attracted the attention of many researchers over the past few years [3].

Fortunately, with the maturation and development of sensor technology and wireless transmission technique over the past few years, more and more relevant technology has been employed from the field of transportation, also the application of this technology has resulted in the developments of Intelligent Transportation Systems (ITS) [4]. Lately, ITS system has gained much recognition, as a result of its efficacy in extremely connected road networks and high demand for road security [5]. In summary, with the deployment of traffic status monitoring device with wireless transmission capacities in current traffic system, the real-time traffic condition data might be transmitted and collected to the related traffic control agency in an accurate and appropriate time, allowing the agency to take further reasonable responsibilities according to the Realtime traffic pattern of the traffic infrastructure for improving targeted traffic control events to minimalize the possibility of traffic congestion [6]. Consequently, a massive amount of ITS-based models was recently introduced and significant research efforts have been dedicated to ITS-based methods. Figure 1 illustrates the outline of ITS communication.

As an essential portion of ITS, an efficient and accurate road traffic predictive method offers accurate and continuous road data according to the previous road condition [7]. This kind of data could be helpful for the ITS applications including vehicular cloud (VC), traffic light control, traffic congestion control, and so on [8]. Now, one of the challenges in maintaining and implementing a VC lies from the computation of accessible redundant vehicular resources on provided road segments, in order to better define the feasible workloads of the cloud architecture. But, the on-road source is largely collected in the higher mobility vehicles on the urban/highway road that makes great importance to the cloud architecture for describing how various vehicles would be on the provided road segment [9]. To address that, the traffic predictive method would offer very consistent traffic volume to the VC based on the past traffic patterns and the spatial relationship on the entire road network. This would

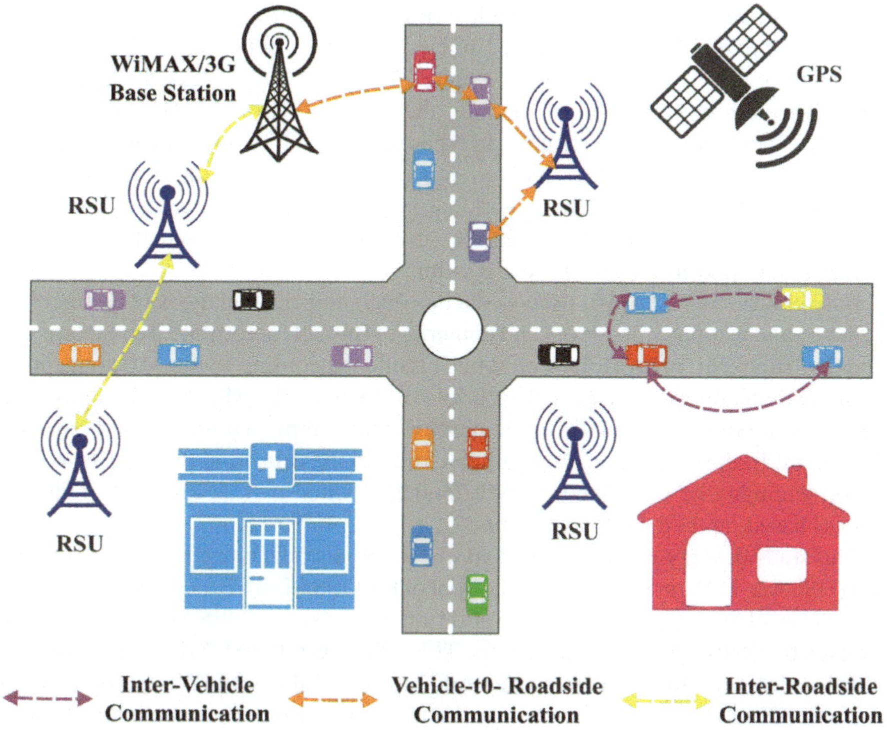

Fig. 1 Overview of ITS communication

provide the VC opportunity for simulating the possible computing and capacity future workloads [10]. As abovementioned, the traffic pattern is forecast based the past traffic flow. In other words, the communication among distinct traffic flows over time would eventually describe the traffic volumes in the future [11]. Indeed, the traffic pattern can be influenced by the different kinds of traffic flow. Generally, the regularities and seasonal trends are influenced by long-term traffic features. In the meantime, the non-seasonal and nonlinear features that are uncertain are influenced by short-term traffic flow.

Meena et al. [12] proposed soft computing, ML, genetic, and DL methods to analyze the big-data for the transport infrastructure with less difficulty. As well, Image Processing algorithm is included in traffic sign detection that ultimately assists in the accurate training of autonomous vehicles. Li and Xu [13] employ ML method for designing different kinds of pattern classifiers, including SVR, Adaboost, SVM, and RF method, to categorize vehicles. The SVR method depends on the fundamental principles of SVM and generalizes to the regression problems. This study presents a short-term traffic pattern predictive method dependent upon SVR and enhances SVM parameters to develop an optimized SVR short-term traffic pattern predictive method.

In Zhu et al. [14], the past vehicle GPS data is utilized for establishing the traffic predictive method. First, the CLIQUE based V-CLIQUE algorithm has been presented for analyzing the previous vehicle GPS information. Next, an ANN based predictive method is introduced. Lastly, the ANN-based weighted shortest path method, A-Dijkstra was presented. In Shao et al. [15] an incremental learning-based CNN-LTSM method, IL-TFNet, is presented for traffic pattern predictions. The lightweight CNN based framework has been developed for processing external environment and spatiotemporal features concurrently to enhance the predictive efficiency and performances of the algorithm.

Boukerche and Wang [16] proposed a thorough and clear analysis of distinct ML methods, and examine the merits and demerits of this ML algorithm. To perform this model, distinct ML algorithms would be classified according to the ML concept. In all the classifications, they provide a brief overview of the ML model and aims at the certain modifications made to the method while employed to distinct predictive challenges. In the meantime, they compare distinct classification that assists to have a short summary of what kinds of ML model is better at what kinds of predictive process based on their exclusive feature.

This study designs an optimal stacked sparse autoencoder based traffic flow prediction (OSSAE-TFP) model for ITS. The goal of the OSSAE-TFP technique is to determine the level of traffic flow in ITS. In addition, the presented OSSAE-TFP technique involves the traffic and weather data for TFP. Moreover, the SSAE based prediction model is designed for forecasting the traffic flow and the optimal hyperparameters of the SSAE model can be adjusted by the use of water wave optimization (WWO) technique. To showcase the enhanced predictive outcome of the OSSAE-TFP technique, a wide range of simulations was carried out on benchmark datasets and the results portrayed the supremacy of the OSSAE-TFP technique over the recent state of art methods.

2 The Proposed Model

In this study, a novel OSSAE-TFP technique is presented to determine the traffic flow in the ITS. The proposed OSSAE-TFP technique is comprised of distinct stages of operations like data collection, SSAE based prediction, and WWO based parameter optimization. TFP method utilizes present weather and traffic-based parameter for estimating the output pattern in subsequent time slices. The result y of the prediction method can be determined in the following

$$y = f(X_1, X_2, X_3, \cdots, X_{10}, X_{11}, X_{12}) \tag{1}$$

Assume that the TF information isn't constrained to the regularity and influenced by the weather condition, the input parameter needs to comprise outside weather conditions. Now, X_i signifies the dataset on the counting flow x_i^{flow}, time slice i, vector depiction of embedding $x^{\text{embedding}}$, weather variables and $x_i^{\text{weather condition}}$ time expression x_i^{timecode}, X_i can be determined as

$$X_i = \left[x^{\text{flow}}, x^{\text{timecode}}, x^{\text{embedding}}, x^{\text{weather condition}}\right], \tag{2}$$

whereas y^{flow} indicates the flow prediction depending upon the traffic series information and $y^{\text{weather \& time}}$ implies the flow predictions depending upon the time periodicity and weather. From the viewpoint of decision level data fusion, the end flow predictive values are the fusion value of 2 decisions, thus the result y of the combined method could be determined in the following

$$y = f_{\text{fusion}}\left(y^{\text{flow}}, y^{\text{weather \& time}}\right). \tag{3}$$

For multistep predictions, y is determined by y_i, and i imply the step size.

2.1 *Data Used*

The traffic datasets of metro freeways in the Twin Cities are utilized. The real dataset is collected at a 30 s period of around 4,500 loop detectors. In the preprocessing phase, the information gets preprocessed in form of table with 5mts period. Furthermore, time similarity measures can be utilized for correcting omissions and errors. To disclose the periodicity of traffic information under weather conflicts, a time-flow relation can be derived. Also, the trained information is separated into non-working and working days, and the average flow in all the time slices is evaluated. The time flow relation for the time slice is given as follows

$$x_t^{\text{timecode}} = \frac{1}{n}\sum_{j=0}^{n} x_{i,j}^{\text{flow}}, \tag{4}$$

whereas $x_{i,j}^{\text{flow}}$ implies the flow of time slice i on day j.

Then, the weather datasets are gathered and utilized 1-hot coding method for handling the non-numerical parameter. In addition, an embedding components are implemented for extracting the expression of high-dimensional information of weather types. Besides, the embedding vector of weather types are given as follows

$$x^{\text{embedding}} = f^{\text{embedding}}\left(x^{\text{one-hot}}\right), \tag{5}$$

Now $x^{\text{embedding}}$ means the trained embedded vector of weather type, whereas x^{one} denotes 1-hot expression. For parameter selection associated with TF, Pearson correlation coefficients ρ is defined by Eq. (6) is evaluated, in which X & Y signifies a set of 2 target variables.

$$\rho_{X,Y} = \frac{cov(X, Y)}{\sigma X \sigma Y} = \frac{E[(X - \mu X)(Y - \mu Y)]}{\sigma X \sigma Y}. \tag{6}$$

For extracting the other weather parameters, the PCA method is applied to the data fusion method at the feature level. The actual matrix A of weather parameter is represented in the following:

$$A = \begin{pmatrix} x_1^{\mathrm{DB}} & x_1^{\mathrm{RH}} & x_1^{\mathrm{Vis}} & x_1^{\mathrm{WB}} & x_1^{\mathrm{WS}} \\ x_2^{\mathrm{DB}} & x_2^{\mathrm{RH}} & x_2^{\mathrm{Vis}} & x_2^{\mathrm{WB}} & x_2^{\mathrm{WS}} \\ \cdots & \cdots & & \cdots\cdots\cdots & \\ x_{n-1}^{\mathrm{DB}} & x_{n-1}^{\mathrm{RH}} & x_{n-1}^{\mathrm{Vis}} & x_{n-1}^{\mathrm{WB}} & x_{n-1}^{\mathrm{WS}} \\ x_n^{\mathrm{DB}} & x_n^{\mathrm{RH}} & x_n^{\mathrm{Vis}} & x_n^{\mathrm{WB}} & x_n^{\mathrm{WS}} \end{pmatrix}, \tag{7}$$

Let x^{DB}, x^{RH}, x^{Vis}, x^{WB}, and x^{WS} be the designated parameters. When the PCA is applied, the resulting matrix P is generated as follows

$$P = f_{PCA}(A) = \begin{pmatrix} x_1^{pca} \\ x_2^{pca} \\ \cdots \\ x_{\mathrm{n}-1}^{pca} \\ x_{\mathrm{n}}^{pca} \end{pmatrix}, \tag{8}$$

Here, x^{pca} represent the fusion value of the chosen weather parameters analyzed by PCA.

2.2 SSAE Based Prediction Model

Primarily, the SSAE based prediction model is designed to forecast the traffic flow in the ITS. The AE was symmetrical NN which removes the feature with minimal reconstructing error. But, the AE has prone to gradient loss or gradient explosion under the trained method because of their several hidden layer frameworks. For solving this challenge, Hinton et al. [17] projected a point of view is named 'pre-train' that separates a difficult network as to stacked sub-network. The trained failure was avoided as network parameters of all the layers are allocated particular values, before arbitrary initialized. However, the stacked sub-network offers minimal trained performance and generalized capability because of the effortlessness of single-hidden layer framework and complexity under the parameter chosen. It can be complex for dealing with diverse, huge, and difficult flame images gathered in various combustion situations to the single-hidden layer framework. For overcoming the above-mentioned restrictions, the SSAE manner was presented dependent upon two-level networks with 5 hidden layers under all the networks. During the 1st-level network, an input layer X was mapped as to hidden layer h_3 (named as lower-level hidden feature). Afterward,

h_3 signifies the mapped back as to reconstructing layer, X_{rec}. During the 2nd-level network, the hidden feature vector h_3 was changed as to input variable for acquiring h_6 (superior-level hidden feature).

During the SSAE, the feature learned procedure follows sequence of functions namely denoising, convolutional, batch normalization (BN), activation, and pooling. An outline of these functions is explained as Denoising: For obtaining the robust as well as representative learning feature of flame images, denoising AE learning technique was utilized by more various noises with input signals. But, distinct kinds of corruption procedures are regarded like mask noises, salt-and-pepper, and Gaussian. During this analysis, the white Gaussian noise was regarded, for instance, the corrupting type X_n has gained with set corrupting ratio to input X, as demonstrated under:

$$I_{x_n} = I_X + \varphi \varsigma \tag{9}$$

where I_X and I_{x_n} refers the pixel intensity of novel as well as noisy images correspondingly; φ indicates the corruption ratio; ς denotes the normal distribution arbitrary variable from the range of -2.576 to 2.576.

Activation function. The ReLU was utilized as activation function of hidden neurons that is determined as [18]:

$$y(\gamma) = \max(0, \gamma) \tag{10}$$

The ReLU was unsaturated piecewise linear operation that is quicker than saturating non-linear operations like Sigmoid as well as TanH. Especially, the sigmoid operation was utilized from the 3rd decoded for ensuring which intensity range of reconstructing layer X_{rec} refers the stable with input layer X that is determined as:

$$y(\gamma) = \frac{1}{1 + e^{-\gamma}}, \tag{11}$$

Pooling and upsampling: The pooling function was implemented for reducing the parameter of network. During this analysis, $P(r \times r + t)$ signifies the pooling layer which reduces the feature mapping by choosing a higher value with $r \times r$ change kernel and step of t. The pooling function was helpful for improving the translation invariance that is written as:

$$\delta_{i,j}^{k} = \max_{0 \le n \le p}\{\gamma_{i \cdot r+t, j \cdot r+t}^{k}\} \tag{12}$$

where $\gamma_{i,j}^{k}$ and $\delta_{i,j}^{k}$ stands for the value of place (i, j) from the kth feature map of input as well as output. During the upsampling layer, bilinear interpolation technique was utilized for achieving the development of feature dimensional that is considered as reversal of pooling function.

Sparse penalty term: While the reconstructing layer of fundamental AE is capable of restoring the input layer appropriately, it can be feasible that network easily copies data in the input-hidden layers. During this manner, the extracted feature was redundant and not enough for more consumption. For avoiding this, the sparse penalties term has further under the SSAE technique that essentially works on hidden layer (h_3 and h_6) for controlling the amount of "active" neurons. But the result of neurons are nearby one, the neuron was regarded as "active", then, it can be "in-active". Generally, it can be superior to avoid the neuron of hidden layer "inactive" and so the learning features are of constraints before simply frequent the input. Let $s_{ij}(i \in (1, E), j \in (1, F))$ defines the activation of hidden neuron j, in which E demonstrates the amount of images under the trained dataset and F signifies the amount of neurons under the hidden layer. Afterward, the average activation of all hidden neurons j is computed as:

$$p_j = \frac{1}{E}\sum_{i=1}^{E} s_{ij} \tag{13}$$

where p_j refers the expecting that nearby 0, significance that neuron of hidden layers is frequently inactive". The penalty term $P_{penalty}$ was enhanced to lose function that penalizes p_j once it deviate importantly in the sparse target p_{target}. The penalty term $P_{penalty}$ was determined as:

$$P_{penalty} = \sum_{j=1}^{F} KL(p_{target} \| p_j) \tag{14}$$

where $KL(p_{target} \| p_j)$ implies the KullbackLeibler divergence (KL divergence) that performs as sparsity constraint and is estimated as:

$$KL(p_{target} \| p_j) = p_{target} \log \frac{p_{target}}{p_j} + (1 - p_{target}) \log \frac{1 - p_{target}}{1 - p_j} \tag{15}$$

When $p_{target} = p_j$, $KL(p_{target} \| p_j) = 0$. Else, the $KL(p_{target} \| p_j)$ improves monotonically as p_j differs in p_{target}.

2.3 WWO Based Hyperparameter Optimization

At this stage, the parameters involved in the SSAE model are adjusted by the use of WWO algorithm. WWO technique was simulated as shallow water wave (WW) model and established by Zheng [19], where all individuals from the population were analogous to "water wave" objects with wave height h and wavelength. Without dropping generalization, supposing have a maximized issue F and their main purpose was f, in which practical issues F are compared with the shallow WW technique; the equivalent relative has demonstrated as:

$$\begin{array}{ccc} \text{Practical issue } F & & \text{Shallow WW manner} \\ \text{The search space of } F & \xrightarrow{\text{analogous to}} & \text{Seabed area} \\ \text{All solutions of } F & \xrightarrow{\text{analogous to}} & \text{A WW objection} \\ \text{The fitness of all solutions} & \rightarrow & \text{It can be inversely proportional to vertical distance to seabed.} \end{array} \quad (16)$$

If the populations were initializing, to all waves, the wave height h was fixed to constants h_{maxi} and wavelength λ was usually set to 0.5. The fitness value of all WWs are inversely proportional to vertical distance to the seabed; it is recognized that under the seabed nearer the WW fitness value was superior, the wave height was greater, and the wavelength was reduced. Under the method of optimized problem solving, search generally from the resultant space with inspired the propagation, breaking, and refraction function of WWs.

Propagation

In WWO, every WW has that propagating if all the generations. It can be considered that novel WW is u, u' refers the novel wave make by propagating operators, the dimensional of maximal value function F is D, the propagating function was relocated, and all dimensions of novel WW u is offered as:

$$u'(d) = u(d) + rand(-1, 1) \cdot \lambda L(d), \quad (17)$$

where $d \in D$, $rand(-1, 1)$ was employed for adjusting the propagating phase that is uniformly distributed arbitrary number set from -1 and 1, and $L(d)$ refers the length of dth dimensional of search space. When the length of $L(d)$ is slower than the length of dth dimensional of search space, a novel place is reset arbitrarily as

$$L(d) = lb(d) + rand() * (ub(d) - lb(d)), \quad (18)$$

where $lb(d)$ and $ub(d)$ stands for the lower as well as upper bounds of dth dimensional of search space and $rand()$ signifies the arbitrary number from the range zero and one. Figure 2 depicts the flowchart of WWO technique.

Then propagate, it can be estimated fitness of u'; if $f(u') > f(u)$, u' before u from the population [20], in the meantime the wave height of u' is reset to h_{maxi}; also, u constant, and for simulating energy dissipation of wave under the method of propagating, their height was reduced by 1.

It can be natural phenomenon that once waves travel in deep water for shallow water, their wave height improves and their wavelength reduces. During the bid for simulating this phenomenon, WWO utilizes the direction in that the wavelength of all waves are upgraded then all the generations as:

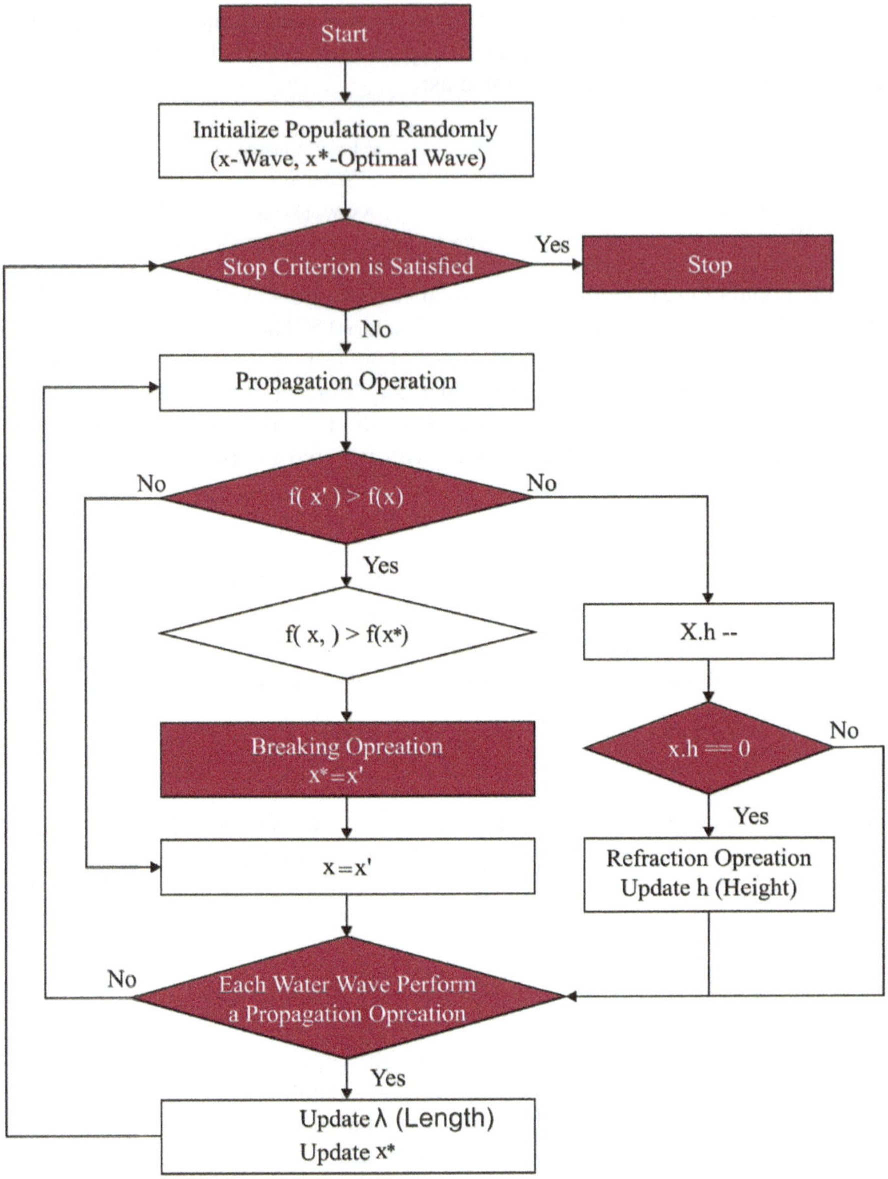

Fig. 2 Flowchart of WWO

$$\lambda = \lambda * \alpha^{-(f(u) - f_{\text{mini}} + \varepsilon)/(f_{\text{maxi}} - f_{\text{mini}} + \varepsilon)}, \tag{19}$$

where α refers the control parameter is known as wavelength reduction coefficients, f_{maxi} and f_{mini} signifies the maximal as well as minimal fitness values amongst the present populations correspondingly, and ε stands for the very low positive constants for avoiding division-by-zero.

Breaking

During the WW technique, with energy of WW improving continuously, crest develops further often steep, and the wave breaks as to sequence of solitary waves if their velocity of crest outperforms the wave celerity. Then propagate, WWO only carries out breaking on wave x that is novel optimum solutions that are utilized for improving the diversity of population. The comprehensive procedure is as follows: initial of all, it can be chosen arbitrarily k dimensional (where k implies the arbitrary number amongst one and existing number k_{maxi}) and implement functions on all chosen dimensional of novel wave u for generating all the dimensional of solitary wave u as:

$$u'(d) = u(d) + N(0, 1) \cdot \beta L(d), \tag{20}$$

where $N(0, 1)$ defines the Gaussian arbitrary number with mean zero and SD one and β demonstrated the breaking coefficients. When the solitary wave with optimum fitness was superior to u^*, u' refers the chosen rather than; then, u^* continued.

Refraction

In WWO, the refracting function only carries out on wave x whose height reduces 0 for avoiding search stagnation that inspires the phenomenon which wave ray could not perpendicular to isobath. With refracting, under the way which arbitrary number centered halfway amongst the novel places and u^* for calculating all dimensional of novel wave, the particulars are as follows:

$$u'(d) = N\left(\frac{(u^*(d) + u(d))}{2}, \frac{|u^*(d) - u(d)|}{2}\right). \tag{21}$$

Afterward refracting, the wave height of u' is also reset to h_{maxi}; in the meantime, their wavelength was upgraded as:

$$\lambda = \lambda \frac{f(u)}{f(u)}. \tag{22}$$

In order to summarize, the function of propagating operator is for making the maximum fitness wave exploiting lesser region and the minimum fitness wave exploring higher region, the breaking operators improve the local explore amongst the capable optimal wave, and the refracting function utilizes avoided explore stagnation and so decreases premature convergences.

3 Experimental Validation

This section inspects the prediction performance of the OSSAE-TFP technique on the test data.

Table 1 investigates the performance validation of the OSSAE-TFP technique interms of MAPE, MSE, and RMSE.

Figure 3 demonstrates the MAPE analysis of the OSSAE-TFP technique with other techniques on the test data. The figure shows that the RBF + PIM, RBF + PCC + PIM, and RBF + PCC + PCA + BP techniques have obtained higher MAPE

Table 1 Performance comparison of OSSAE-TFP technique with recent techniques on weather data

Methods	MAPE (%)	MSE	RMSE
RBF + PIM	21.441	404.123	20.103
RBF + PCC + PIM	21.070	398.337	19.958
RBF + PCA + PIM	19.480	389.533	19.737
RBF + PCC + PCA + PIM	19.171	385.362	19.631
RBF + PCC + PCA + BP	21.909	517.920	22.758
OSSAE-TFP technique	17.565	308.356	17.560

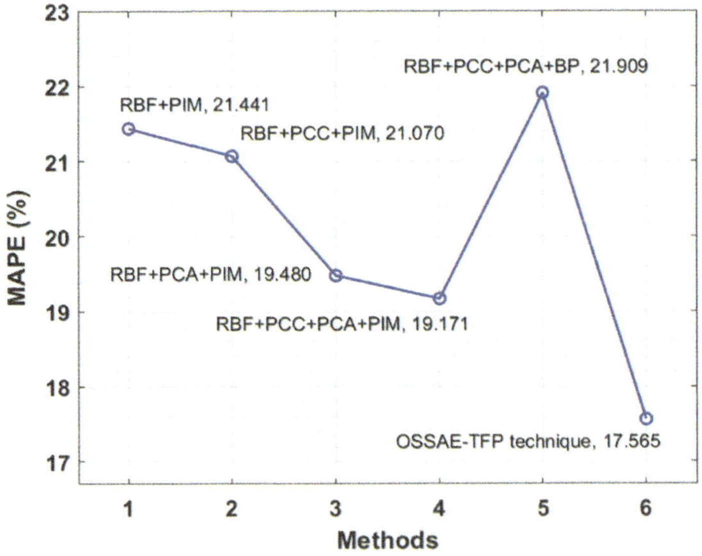

Fig. 3 MAPE analysis of OSSAE-TFP model with existing approaches

of 21.441, 21.070, and 21.909 respectively. In addition, the RBF + PCA + PIM and RBF + PCC + PCA + PIM techniques have obtained moderate MAPE of 19.480 and 19.171% respectively. However, the proposed OSSAE-TFP technique has accomplished improved performance with the MAPE of 17.565%.

Figure 4 depicts the RMSE analysis of the OSSAE-TFP manner with other methods on the test data. The figure outperformed that the RBF + PIM, RBF + PCC + PIM, and RBF + PCC + PCA + BP techniques have obtained superior RMSE of 20.103, 19.958, and 22.758 correspondingly. Besides, the RBF + PCA + PIM and RBF + PCC + PCA + PIM techniques have gained moderate RMSE of 19.737 and 19.631 respectively. But, the projected OSSAE-TFP technique has accomplished increased efficiency with the RMSE of 17.560.

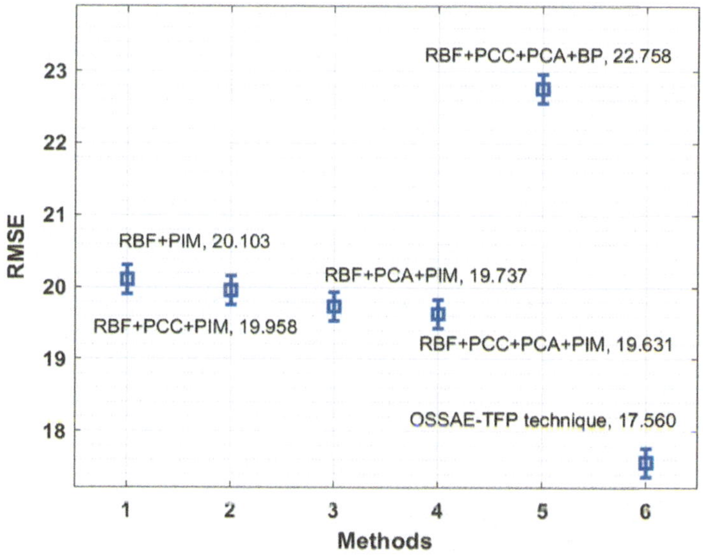

Fig. 4 RMSE analysis of OSSAE-TFP model with existing manners

Table 2 Comparison study of AITFP-WC method with existing techniques

Methods	MAPE (%)	MSE	RMSE
LSTM	11.692	166.494	12.903
GRU	12.238	167.223	12.931
S.LSTM	11.125	156.164	12.497
S.GRU	11.279	158.288	12.581
SAE	11.226	162.112	12.732
Fusion-ANN	11.185	151.708	12.317
SAERBF	10.965	151.825	12.322
OSSAE-TFP technique	9.568	100.384	10.019

Table 2 provides a comprehensive comparative study of the OSSAE-TFP technique with recent methods. The figure shows that the GRU model has showcased worse outcomes than the other ones. Followed by, the LSTM, S.LSTM, S.GRU. SAE and Fusion-ANN models have reached moderate performance. Though the SAERBF technique has accomplished reasonable performance with the MAPE, MSE, and RMSE of 10.965%, 151.825, and 12.322, the proposed OSSAE-TFP technique has resulted in maximum outcome with the MAPE, MSE, and RMSE of 9.568%, 100.384, and 10.019.

A detailed MAPE analysis of the OSSAE-TFP technique with other techniques under varying time duration takes place in Table 3 and Fig. 5. The results show that the OSSAE-TFP technique has accomplished improved outcomes with the minimal MAPE under every time duration. For instance, with 5 min, the OSSAE-TFP technique has gained a lower MAPE of 9.57 whereas the LSTM, GRU, S.LSTM, S.GRU,

Table 3 Result analysis of AITFP-WC model interms of MAPE

Methods	5 min	10 min	15 min	20 min	25 min	30 min
LSTM	11.69	14.63	16.84	18.17	19.77	20.63
GRU	12.24	15.15	16.87	18.72	19.52	20.15
S.LSTM	11.13	14.48	16.07	16.56	18.00	19.48
S.GRU	11.28	13.96	14.83	17.02	17.91	18.96
SAE	11.23	13.86	15.90	16.98	18.15	19.86
FUSION-ANN	11.19	12.94	14.68	15.79	16.00	18.94
SAERBF	10.97	13.14	14.41	14.71	16.05	18.14
OSSAE-TFP technique	9.57	12.42	14.00	14.22	15.68	16.64

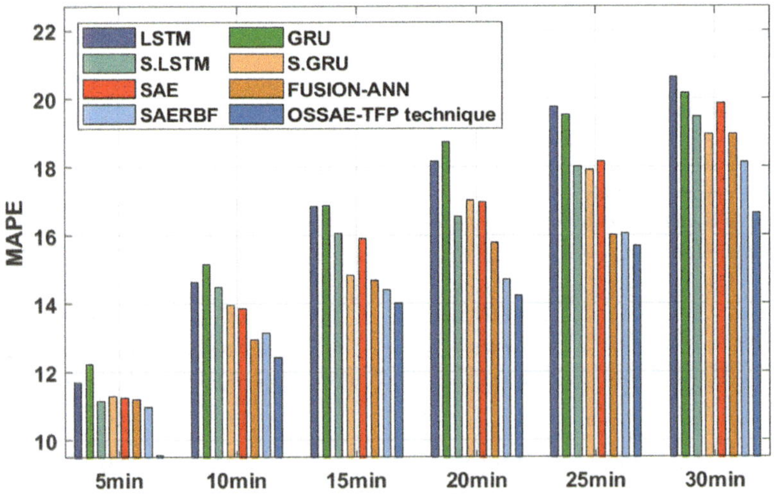

Fig. 5 MAPE analysis of OSSAE-TFP technique under varying time duration

SAE, FUSION-ANN, and SAERBF techniques have accomplished higher MAPE of 11.69, 12.24, 11.13, 11.28, 11.23, 11.19, and 10.97 respectively.

At the same time, with 10 min, the OSSAE-TFP manner has reached a minimum MAPE of 12.42 whereas the LSTM, GRU, S.LSTM, S.GRU, SAE, FUSION-ANN, and SAERBF techniques have accomplished superior MAPE of 14.63, 15.15, 14.48, 13.96, 13.86, 12.94, and 13.14 correspondingly. Moreover, with 20 min, the OSSAE-TFP technique has gained a lower MAPE of 14.22 whereas the LSTM, GRU, S.LSTM, S.GRU, SAE, FUSION-ANN, and SAERBF algorithms have accomplished higher MAPE of 18.17, 18.72, 16.56, 17.02, 16.98, 15.79, and 14.71 correspondingly. Furthermore, with 30 min, the OSSAE-TFP approach has achieved a decreased MAPE of 16.64 whereas the LSTM, GRU, S.LSTM, S.GRU, SAE, FUSION-ANN, and SAERBF systems have accomplished higher MAPE of 20.63, 20.15, 19.48, 18.96, 19.86, 18.94, and 18.14 correspondingly.

A brief RMSE analysis of the OSSAE-TFP manner with other algorithms under varying time duration takes place in Table 4 and Fig. 6. The outcomes outperformed that the OSSAE-TFP technique has accomplished increased outcome with the minimal RMSE under every time duration. For instance, with 5 min, the OSSAE-TFP approach has attained a reduced RMSE of 10.02 whereas the LSTM, GRU, S.LSTM, S.GRU, SAE, FUSION-ANN, and SAERBF methodologies have accomplished higher RMSE of 12.90, 12.93, 12.50, 12.58, 12.73, 12.32, and 12.32 correspondingly. Simultaneously, with 10 min, the OSSAE-TFP technique has gained a lower RMSE of 13.49 whereas the LSTM, GRU, S.LSTM, S.GRU, SAE, FUSION-ANN, and SAERBF systems have accomplished higher RMSE of 14.74, 14.78, 14.19, 14.02, 14.67, 13.81, and 13.90 correspondingly. In addition, with 20 min, the OSSAE-TFP technique has gained a lower RMSE of 14.97 whereas the LSTM, GRU, S.LSTM, S.GRU, SAE, FUSION-ANN, and SAERBF techniques have accomplished increased RMSE of 17.60, 17.44, 16.40, 17.01, 17.63, 15.59, and 15.46 respectively. At last, with 30 min, the OSSAE-TFP technique has gained a decreased RMSE of 17.94 whereas the LSTM, GRU, S.LSTM, S.GRU, SAE, FUSION-ANN, and SAERBF manners have accomplished maximal RMSE of 20.59, 21.30, 19.27, 19.53, 21.34, 18.42, and 18.39 correspondingly.

Table 4 Result analysis of AITFP-WC model interms of RMSE

Methods	5 min	10 min	15 min	20 min	25 min	30 min
LSTM	12.90	14.74	16.55	17.60	19.09	20.59
GRU	12.93	14.78	16.56	17.44	19.74	21.30
S.LSTM	12.50	14.19	15.14	16.40	17.96	19.27
S.GRU	12.58	14.02	15.17	17.01	18.14	19.53
SAE	12.73	14.67	15.86	17.63	19.33	21.34
FUSION-ANN	12.32	13.81	14.67	15.59	17.12	18.42
SAERBF	12.32	13.90	14.57	15.46	17.19	18.39
OSSAE-TFP technique	10.02	13.49	14.08	14.97	16.75	17.94

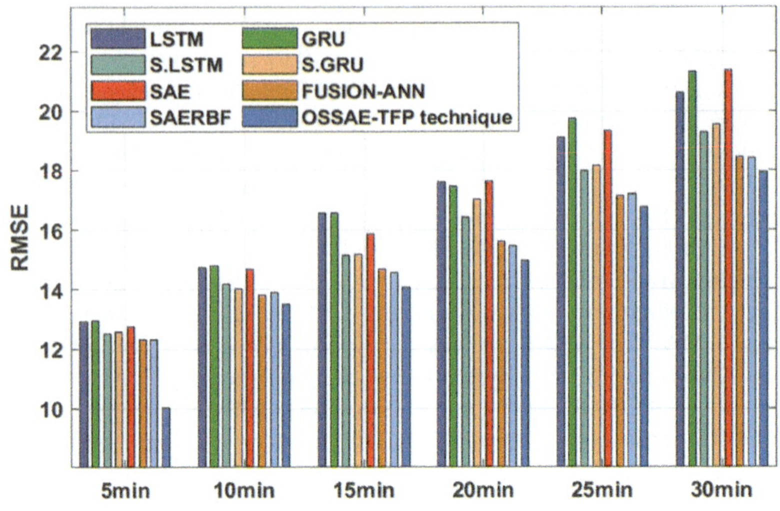

Fig. 6 RMSE analysis of OSSAE-TFP technique under varying time duration

4 Conclusion

In this study, a novel OSSAE-TFP approach was projected to determine the traffic flow in the ITS. The proposed OSSAE-TFP technique comprised of distinct phases of operations like data collection, SSAE based prediction, and WWO based parameter optimization. Primarily, the SSAE based prediction model is designed to forecast the traffic flow in the ITS. At the next stage, the parameters involved in the SSAE model are adjusted by the use of WWO algorithm. In order to showcase the enhanced predictive outcome of the OSSAE-TFP technique, a wide range of simulations was carried out on benchmark datasets and the results portrayed the supremacy of the OSSAE-TFP technique over the recent state of art methods. Therefore, the proposed OSSAE-TFP technique can be utilized as an effective tool to predict the flow of traffic in the ITS. In future, advanced DL models can be used in place of SSAE model to improve the predictive outcome.

Conflict of Interest The authors declare that they have no conflict of interest. The manuscript was written through contributions of all authors. All authors have given approval to the final version of the manuscript.

Data Availability Statement Data sharing not applicable to this article as no datasets were generated during the current study.

References

1. Boukerche, A., Tao, Y., Sun, P.: Artificial intelligence-based vehicular traffic flow prediction methods for supporting intelligent transportation systems. Comput. Netw. **182**, 107484 (2020)
2. Sun, P., Boukerche, A.: AI-assisted data dissemination methods for supporting intelligent transportation systems. Internet Technol. Lett. e169 (2020)
3. Zhang, J., Wang, F.-Y., Wang, K., Lin, W.-H., Xu, X., Chen, C.: Parallel control and management for intelligent transportation systems: concepts, architectures, and applications. IEEE Trans. Intell. Transp. Syst. **11**(3), 630–638 (2010)
4. Wang, F.-Y.: Toward a revolution in transportation operations: AI for complex systems. IEEE Intell. Syst. **23**(6), 8–13 (2008)
5. Iordanidou, G., Papamichail, I., Roncoli, C., Papageorgiou, M.: Feedback-based integrated motorway traffic flow control with delay balancing. IEEE Trans. Intell. Transp. Syst. **18**(9), 2319–2329 (2017)
6. Sun, P., AlJeri, N., Boukerche, A.: DACON: a novel traffic prediction and datahighway-assisted content delivery protocol for intelligent vehicular networks. IEEE Trans. Sustain. Comput. 1 (2020)
7. Alipour, B., Tonetto, L., Ketabi R., Yi Ding, A., Ott, J., Helmy A.: Where are you going next? A practical multi-dimensional look at mobility prediction. In: ACM MSWIM, 2019, pp. 5–12
8. Domingues, A.C.S.A., Silva, F.A., Loureiro, A.A.F.: On the analysis of users' behavior based on mobile phone apps. In: ACM MobiWac, 2019, pp. 25–32
9. Meneguette, R.I., Nakamura, L.H.: A flow control policy based on the class of applications of the vehicular networks. In: ACM MobiWac, 2017, pp. 137–144
10. Zhang, T., Grande, R.E.D., Boukerche, A.: Vehicular cloud: Stochastic analysis of computing resources in a road segment. In: Proceedings of the ACM Symposium on Performance Evaluation of Wireless Ad Hoc, Sensor, & Ubiquitous Networks, 2015, pp. 9–16
11. Sun, P., AlJeri, N., Boukerche, A.: A fast vehicular traffic flow prediction scheme based on Fourier and wavelet analysis. In: IEEE Global Communications Conference, IEEE, 2018, pp. 1–6.
12. Meena, G., Sharma, D., Mahrishi, M. (2020, February). Traffic prediction for intelligent transportation system using machine learning. In: 2020 3rd International Conference on Emerging Technologies in Computer Engineering: Machine Learning and Internet of Things (ICETCE), pp. 145–148. IEEE
13. Li, C., Xu, P.: Application on traffic flow prediction of machine learning in intelligent transportation. Neural Comput. Appl. **33**(2), 613–624 (2021)
14. Zhu, D., Du, H., Sun, Y., Cao, N.: Research on path planning model based on short-term traffic flow prediction in intelligent transportation system. Sensors **18**(12), 4275 (2018)
15. Shao, Y., Zhao, Y., Yu, F., Zhu, H., Fang, J.: The traffic flow prediction method using the incremental learning-based CNN-LTSM model: the solution of mobile application. In: Mobile Information Systems, 2021 (2021)
16. Boukerche, A., Wang, J.: Machine Learning-based traffic prediction models for intelligent transportation systems. Comput. Netw. **181**, 107530 (2020)
17. Hinton, G., Salakhutdinov, R.: Reducing the dimensionality of data with neural networks. Science **313**(5786), 504–507 (2006)
18. Han, Z., Hossain, M.M., Wang, Y., Li, J., Xu, C. (2020). Combustion stability monitoring through flame imaging and stacked sparse autoencoder based deep neural network. Appl. Energy **259**, 114159
19. Zheng, Y.-J.: Water wave optimization: a new nature-inspired metaheuristic. Comput. Oper. Res. **55**, 1–11 (2015)
20. Wu, X., Zhou, Y., Lu, Y.: Elite opposition-based water wave optimization algorithm for global optimization. In: Mathematical problems in engineering, 2017 (2017).

Leverage Computer Vision for Cost-Effective Learning Paradigm

Ankit kumar, Atul Yadav, Ashutosh Kumar, Amit Kumar, and Achal Kaushik

Abstract In today's world, technologies like virtual and augmented reality grow every second and provide diverse dimensions to various other domains. These technologies are more likely to bring users into the virtual world to help them understand multiple concepts more closely. In fact, in the present era, little kids of the current generation are more curious about the know-how of things in a more visual way. The curiosity among the young kids to understand various concepts requires a better learning environment. The kids may explore the working of the objects in 3D by visualizing the natural movements. But practically, this demands a virtualized environment for them to understand these things and that too in a safe environment. Now, this is somehow getting possible due to the latest advancement within these technologies which opens space for exploration for various domains. Recently Google has developed ARCore, an SDK that allows the development of augmented reality-based applications. In this virtualized environment, the user can immerse themselves even when they are not physically present. In our model, we have facilitated the user to explore different dimensions of the computer vision which are required for understating various concepts in a very cost-effective manner. In our model, we have provided a vision-based interface to the user that replaces peripheral hardware without affecting their performances and facilitates them to use these interfaces embedded with AR to better understand 3D concepts.

Keywords OpenCV · OpenGL · Computer vision · Virtual reality · Augmented reality · ARCore

1 Introduction

Computer Vision (CV), Virtual Reality (VR), Augmented Reality (AR) and related technologies have the potential to improvise on the learning experience. These technologies provide students with an immersive experience of being live within the

---

A. kumar · A. Yadav · A. Kumar · A. Kumar · A. Kaushik (✉)
Bhagwan Parshuram Institute of Technology, Affiliated to GGSIPU, Delhi, India

A. E. Hassanien et al. (eds.), *Virtual and Augmented Reality for Automobile Industry: Innovation Vision and Applications*, Studies in Systems, Decision and Control 412, https://doi.org/10.1007/978-3-030-94102-4_7

environment without being physically present. This virtual environment helps them learn more interactively with deep understanding [25, 44]. An Amalgamation of the natural world with the virtual objects in a digital overlay provides a way of enhancing the understanding of real-world objects. This exposure is possible by adding the information generated by computers. Computer vision will bring the learners close to the complex concepts by using and demonstrating the working models, and this will enhance the overall understanding level [43]. At times, students in their laboratory can perform hazardous science experiments with ease and visualize the complex 3D structures [2, 6].

In rural areas, students are not exposed to technological advancements. Often, it is not cost-effective for them to utilize costly gadgets for learning, playing, and understanding the concepts using costly hardware. Although, the basic concept of advanced technologies like computer vision etc. can be applied in the field of elementary education for their easy grasping and also for secondary school students, who find it quite tough to visualize difficult concepts like atomic structure and fundamentals of quantum physics and their working mechanisms [25, 39].

Computer games are also one of the most famous electronics items and gadgets that are present around us. Users and especially game players find it breadth-taking to use and indulge with the gaming world via joysticks, trackball, buttons or wired gloves [1]. Hence at times, they find it much better and satisfying to interact through natural, body motions, or various kinds of gestures [41]. Smartphones have cemented their place in our daily life, with a lot many activities controlled directly by them. It is no longer a costly affair for the users, and they are also used for various purposes like for communication or making calls and for other academic purposes. But managing several peripheral devices is not convenient while playing or learning and sometimes it is not very cost effective.

There are different methods available to perform the task to convert any smartphone into a wireless mouse or keyboard. One way is by making use of the internet and IP address of the machine the connection can be made and alternatively, control PC via Android phone through Android remote control. This is referred to as casting where the mobile screen can be mapped to the desktop or TV screen and the operation can be performed wirelessly.

In these systems, the application is developed to work on a client–server architecture where the server-side is a PC, and the client-side is a smartphone. This client–server interaction can be handled using wi-fi where commands and actions are translated on both sides in the form of packets. Keypress actions are tapped by the server when pressed on the client and wirelessly the information packets are transferred. Similarly, the mouse click action mapped to the gesture control where client-side gesture specification mapped to the server [41].

These advances provide flexibility to smartphone users to use the existing device for multitasking purposes. Now, with the advent of new technologies, like augmented reality, virtual reality and immersive experience etc. bringing larger than life to the gaming systems like Pokemon-go games etc. At the same time, the hardware complexity and costing are the factors limiting the easy incorporation of these technologies [13].

Various devices like keyboards, mice, joysticks and other peripheral devices are an integral part of the system for their efficient usage [10]. In various instances, we have seen that users are quite dependent on these devices, but our model will provide users freedom from multiple devices. We can control our system from a distance apart and hence will increase efficiency and decrease the dependency. Our concept will also help in decreasing electronic waste as lesser keyboards and mice will be used [42]. Apart from that our concepts will also provide a level playing field to the gamers, learners where they can compete with high performance.

In our work, we envisage the concept of easy cost-effective learning for the students using fewer devices. One of the core advantages of our work is that it establishes the link between virtual reality and augmented reality in a synchronized way and thus bridges the gap between these two recent technologies for better-grasping concepts for the students in a cost-effective manner [7].

In our approach, we have used the simple concept of providing a vision-based interface to replace the input devices used for playing different PC games and learning different concepts. We are proposing a method where users can use any object available for replacing joystick or keyboard controls to play computer games and subsequently provide them with an interface using AR for understanding the 3D concepts.

Our work comprises of two parts, the first one is based on the android app to effectively reduce the hardware dependency and the second on top of it is the use of computer vision technology [12]. In the first part, we have mainly focused on electronic technology utilization for minimizing the e-waste and decreasing the complexity of operations for various proposes like official work, gaming and various applications by simply taking control of the operations by using android applications and smartphones [14, 15]. The android app will replace basic functions like mouse, keyboard, and gaming pads and this will decrease the additional cost on the economic front and complexity on the operational front. This will further improve the domain of the application for our usability.

In this first part of our model, we provide gaming/learning interface usability where any object can be used as an interface for the gaming controls. A simple piece of paper may be used to play the game [51]. To achieve this, we need to provide a mechanism through which object detection and traceability of its motion are mapped. In this approach, the user can play a game or study using an object just like he plays/works with a mouse, keyboard, or gaming. In the second approach adding the concept of AR, we can understand any complex phenomenon with 3D visualization. The main advantages of our model are that it's highly economical as if we are using any object and it can be replaced by a new one if one is damaged and without any costly hardware, we can showcase 3D concepts for better understanding. Hence, the model will also provide domain usability for users by using an economical technique.

The different sections of this paper are organized as follows. In Sect. 2, the literature review showing previous works on similar technologies are discussed. In Sect. 3, we have first discussed our proposed model and subsequently discussed the features of our model. The application and performance comparison of our system is shown in Sects. 4 and 5 offers conclusions and direction for future work.

2 Literature Review

Computer vision is an interdisciplinary field that provides the mechanism with which computers can use to gain high-level understanding from video, images, and interactive text. Hence, we can say that from a technological aspect that OpenCV provides the automation for the mechanism of performing the task that a human visual system can perform [12].

From exhaustive research [3, 18, 23, 36], we have found computers are highly capable are recognizing images and related contents, that is why top companies like Google, Amazon and Facebook are investing high capital into this technologies, various applications sector of OpenCV are banking, healthcare, agriculture, space technologies etc. [28, 40].

Many researchers are working on improvising vision-based user interface systems. In literature, it is found that different authors [9, 27, 30, 51] have used different approaches for hand tracking systems. For tracking systems, some have used grayscale images while some others have used the feature extraction method. In another approach, authors have used the concept of finger tracking to move virtual items and hand posture to control virtual presentations while some use different combinations of fingers like forefinger and thumb for the hand tracking system. In Zhang et al. [51] authors used fingertips as pointer type input devices and even used for drag and drop items. In Lee and Lee [30] authors have developed region-based tracking for movement and pointing based on a fingertip. This computer vision-based mouse performs mouse actions within the view of the camera.

In Kumar et al. [27], Banerjee et al. [9] colour-based mouse actions are performed using a real-time camera. Colour-tapes are used on the fingers to control or colour detection techniques are used for the recognition of the hand gestures. In Murase et al. [37] developed a virtual keyboard for facilitating virtual keys entering using gesture capturing. In this system, only the webcam is required to map the QWERTY keyboard layout for performing similar functions. In Lee and Seo [31] authors have proposed a vision-based virtual joystick where a handheld stick is acting as an interface for the users. The joystick-like movement of the stick is controlled by hand movement tracking using a camera. In Veluchamy et al. [48] the authors have developed interfacing to the computer system with the virtual environment using hand gestures where the system is recognizing the hand gestures through the webcam on the computer and providing a vision-based gesturally controllable interaction system.

Expanding studies on AR have covered a wide variety of techniques and methodologies like motion tracking, light estimation, AR Core environment understanding etc. Several researchers have contributed to the advancements of the AR concept, while initially it was centered around head-mounted displays (HMDs), in recent years advances to handheld AR devices, with state-of-the-art hardware and wearable mobile devices [17].

The application use of AR is not restricted to only games but it is widely used in healthcare [35], education, and other fields [6]. Imparting and improving the education standards have been a central concern for AR developers. The Augmented reality-based education system for early learners enables them to recognize shapes, alphabets, colours using flashcards [33]. Authors [19] have developed a teaching tool based on AR using 3D visualization. The study on mixed-method research evaluates the learning of university students using AR and visual technology and found improved results.

Health issue concerns were also addressed using AR and research on movements of body parts for physiotherapy in an active text drawing for outdoor AR using the concept of scene form API of light estimation [20, 35]. There are other research studies like AR-based video communication with the capability of improvising on user experience with social presence [47], gesture-based interaction with voice feedback for a smartphone [5] and through the looking glass pretend to play with children [8].

Authors [24] proposed face to face collaborative AR games using the basic but efficient concept of 3D modelling in AR and motion tracking. A multiplayer collaborative game for primary school children is developed using the Augmented Reality concept [33]. Strong bonds between participants foster their socialization and communication. It is observed that AR as a concept and technology is only going to improve and be improvised by technologies like AR Core motion tracking, environmental understanding, and virtual reality. In our work we have tried to use and inculcate concepts like motion tracking, 3D modelling, light estimation etc., but with a more efficient and technologically dependent specification of resources and utility considering further needs.

Our model has wide application in the field of learning, gaming, and online avenues especially gaming enthusiasts will find it quite handy as the game controller will help in reducing the gadget dependency of game lovers on various devices like laptops, mice, and joystick etc. Apart from the gaming world, our model can also be used in the education and corporate world where during presentations or learning the concepts, the 3D display and projector can be controlled using your android device and hence will minimize the dependency on various hardware devices.

Our concept also has the potential to modify and elevate the gaming standards due to its usage in the gaming industry and the use of OpenCV and machine learning. Prominently due to the use of OpenGL [36, 49], gaming efficiency can be improved with superior graphics and hence will provide a thrill to game lovers across the globe, also OpenCV uses low-level C++ language which has applications in the gaming industry as it provides the flawless experience for the user due to high-quality standards and hence has the tendency to change the face gaming over mobile, laptop and desktops [18, 23, 34]. Apart from that its implementations involve computer vision framework (OpenCV) and SDL (Simple Direct Media Layer) for efficient gaming and graphics [38]. Using the game controller any gamer with this application can evenly play and compete with other gamers of the gaming fraternity.

3 Proposed Method

We have categorized our model into different stages to facilitate a virtual interface with 3D view concepts for the learning/gaming sub-systems (Fig. 1). In the first stage, we have discussed socket programming which is used to connect two devices for virtualization and after that, we have discussed other stages required for materializing the different concepts of our model. These steps involved the filtration process that will take place to remove the unwanted objects from the frames. In the next step, identification of the object will take place and subsequently the direction of movement as per environment/gaming condition is established and lastly the Plane Detection and Object Placing facility for implementing 3D concepts.

3.1 Socket Programming

Communication of data between two devices over the network must follow some protocols and establish the connection between them. Socket programming is the interface between the two devices which has some information about the devices. Networks are composed of two types of devices: server and client where the clients are those who send the request either for getting resources and other information while the servers are those which receive data from clients and send the response in the form of data [15, 26].

To communicate between clients and servers, both sides have a defined socket which contains the IP (internet protocol) address and port by which other devices will connect, so in our model, we create the server socket on the desktop using the java socket programming which contains different APIs methods for getting the request and sending the response to the client. To create the server socket on the desktop we have used the ServerSocket object which will take the IP and port to open the connections for clients. Following are the steps for socket programming (Fig. 2):

(a) Create a server socket and open the connection for the specified IP address and port
(b) A different thread will create which will be in the event loop for client request
(c) On the client-side, it will create a client socket that has the IP and port of the server
(d) After the connection is established, it will start sending the data to the server

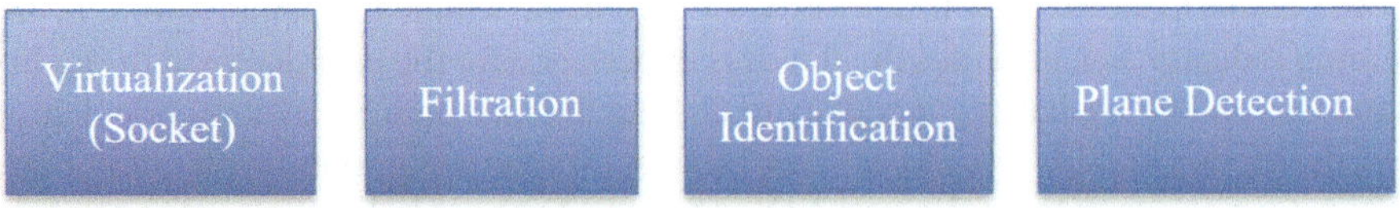

Fig. 1 Implementation steps

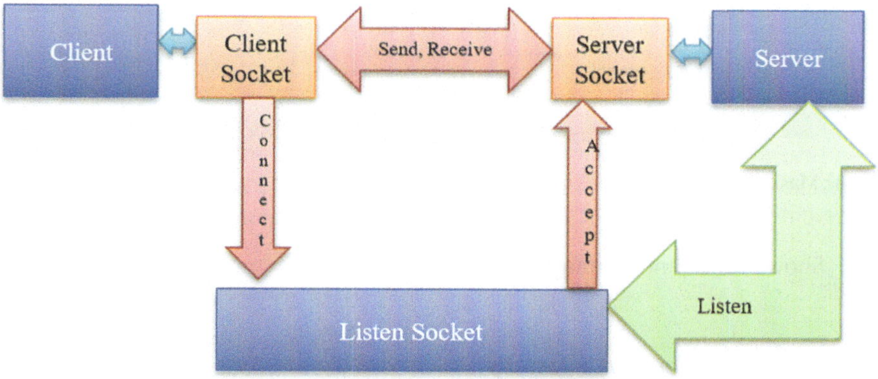

Fig. 2 Socket connectivity

(e) On the server, it will evaluate the data coming from the client-side
(f) In the end, all the connections will be destroyed when all communications are done.

3.2 Filtration Process

All images have different segments, which includes noise or other image structures. The image filtering is applied to reduce the noise, filtering like linear translation-invariant like a gaussian blur, Laplacian filter and Sobel filter are being used in the image filtering (Fig. 3). These techniques provide a different method of filtering like blurring the image, edge detection and feature extraction [21].

In our work, we apply the smoothing technique for image flirting, this is also called blurring which is often used for image processing. A Linear filter is used to filter the image, the weighted sum of input pixels of the image is calculated for the output pixel of the image and stored in the matrix form and the weight of pixels in linear operation is usually called the kernel [21].

Steps for filtering the image:

(a) Getting the image from the front camera
(b) Convert the image from BGR to HSV
(c) Apply gaussian blur

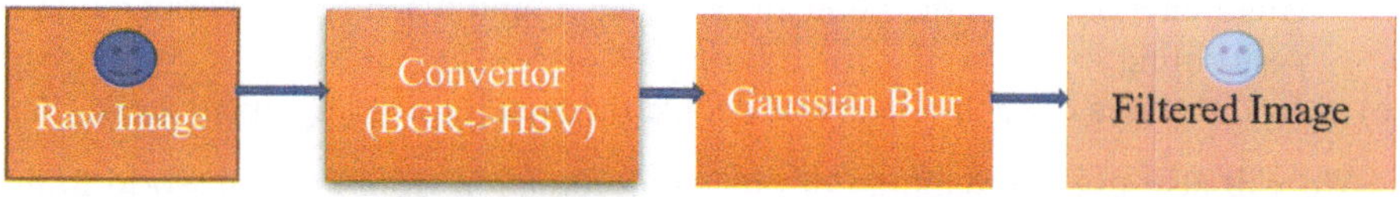

Fig. 3 Filtration step

Fig. 4 Masking step

(d) Show reduced noise image.

3.3 *Object Detection Process*

In this process, a class of objects is identified in the image. This is required to properly isolate the object from the surrounding area. For example, we are using object detection for getting an understanding of the shape which will drive the game and give the information about the object to the program for further operations. We have used different OpenCV functions for detecting the colour by performing the masking operation on the HSV image [16]. The following steps are used for detection:

(a) Open the default camera
(b) Matrix for capturing the image and storing the image in the matrix
(c) Convert the BGR colour image into HSV colour
(d) Providing the HSV value in the range from upper to lower for considering the given colour
(e) Perform the masking on the matrix of the image (Fig. 4).

3.4 *Object Movement Process*

The next step is to detect the movements and for that, a proper selection of a set of motion tracking parameters are required. These parameters are also used for effective performance requirements for surveillance systems and have invariably been used with some method for characterizing video datasets [50]. The following steps are used for motion tracking:

(a) Find Contours where the picture is the image matrix contours and hierarchy are the vector for storing the contours
(b) Hierarchy expresses the number of objects found when contours operation is performing
(c) Find the area of each object and decide the area for filtering the process.

We will also use the mapping of object coordination from the real world to the virtual world (in the computer), each time position of the object and its coordinate

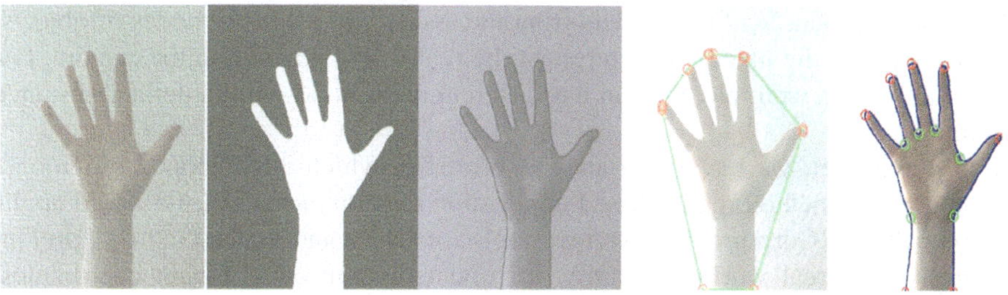

Fig. 5 Morphing steps

will decide the movement of the game either it will move left or right so this is how we make the object movement in our model.

In our application software, we have used the visual studio which provides the integrated development environment and for the computer vision, we have used OpenCV which is an open-source library [43] for developing the computer vision application. Our approach here is first implementing the colour detection [46] in C++ and after getting the data from the colour detection, use that data for implementing the tracking object [4] of that particular colour called the object tracking (Fig. 5).

We determined the detected object movement in different directions which includes finding the contours of the object and how much area shifted from the previous position. We also map the world coordinate of the object into computer screen coordinate so that we can determine its position for performing the game controlling [50]. We have used computer vision for object detection and getting the coordinates of the object below is the sample code.

For Tracking

(a) Get the image from the input
(b) Convert the BGR colour to HSV colour
(c) Define the HSV lower bound and upper bound.
(d) Define the range of the HSV colour and call the function of OpenCV
(e) Call the morp function to morph the object [11].

3.5 Plane Detection and Object Placing

In our approach, we have used the Sceneform fragment, where it defines all the permissions related to the camera and user features that should be present on the phone. Then it starts the augmented reality sessions and Sceneform API provides the listener (Observer pattern) for plane detection. The in-built feature indicates for the user to move their camera to detect the plane view and subsequently allows to place a 3D object in the scene view. It performs the animation on these 3D objects with properties like a rotation around the parent node (sun) and rotation around their own axis [45]. All these animations are performed by the object animation and with the same fragment load the models from the assets and make the nodes of objects. Nodes here are the objects which reflect the real-world position in the camera. For the solar system example, the sun has different child nodes with a defined distance around it.

In our model, we have used an ARCore SDK which is developed by Google to support augmented reality-based application development. For AR-based applications, the ARCore provides various tools for using and understanding different objects of the real world. Primarily, for understanding environment coordinates, which enables the detection of the vertical and horizontal planes, motion tracking which enables the device to understand their respective positions and relatively track them to the real world and the complex object detection, a particular concept of light estimation API which enables the digital objects to appear realistically. The ARCore elements use a set of common AR UI components which is validated with the help of user testing. For example, plain detection puts dots to demarcate the surface and enables the user to move or transform objects by zooming in and out or rotating the objects by hand gestures on the screen.

Also, the ARCore uses 3 key technological concepts to give integrated virtual content with the real world of a camera [22, 29, 32].

(a) Six degrees of freedom enables the device to understand and track the position
(b) Environment understanding enables the device to detect the location and the size of vertical or horizontal surfaces like walls, floors, tables etc.
(c) Light estimation enables the device to estimate the real world's contact and lighting conditions.

The steps for plane detection and object placing are the following:

For Plane tracking

(a) Define Sceneform fragment by defining ID as ArFragment
(b) Listen for plane detection on the fragment
(c) Update the frame when we detect the plane.

For object placing

(a) Create a node at a specific position from the anchor node
(b) Create a child node of the anchor node and set its local position.
(c) Now create a transformable node that has properties of rotation and scaling and set it as a child node.
(d) Define the path of the 3D model in the transformable node.
(e) Perform the animation on the node by Object animation and select the nodes.

4 Application and Features

Our model has wider applications prominently in the field of gaming industry where game users will find it suitable and exciting due to its gadget independent feature where our android device can be used to control our computer reducing the use of keyboards and mice. The model has its applicability in the education domain where students can learn the concept efficiently using the latest technologies more like giving them an immersive environment for an easily understandable view of concepts.

The model can be used to decrease the e-waste generated by various gadgets and devices which were earlier inseparable from the system. It will help authorities to tackle the menace of electronic waste generation. One more application of our model will be in saving costly devices like mice, keyboards and joysticks from children's reach and protecting the damage to them, by using a singular device or even a piece of paper for gaming.

Our model will make the whole process less time consuming and more efficient to utilize the technology. In this multi-user environment, where at the same two or more users can connect to the desktop for controlling the desktop and make use of different points for using the android application. Further, a model based on the AR provides the visualization of those objects which are rare to see or having complex architecture, for example, we provide the experience of the solar system (Fig. 6) which will provide the information of our solar system and how the planets are revolving around the sun and also how these planets rotate on their axis with control interaction [45].

Many a time student only learn theoretical concepts about the objects like animals they never get a chance to closely look at those objects and see how they look, behaves etc. but to overcome this our model provides various 3D models which children can

Fig. 6 Solar system

use to examine and understand different properties by a simple click on the surface where he wants to place the objects.

We also provide the feature to measure the length of an object in real life (ruler) where students can use to find the length of any object placed in real life, it only requires marking the objects at both the ends and application will show the length of the component, this would help students to learn about distance and how to measure the objects (Fig. 7).

From Fig. 7, we have shown the distance calculation between different objects. In this process, the user can put two objects in the real world and using the application we can evaluate the distance between the objects. This helps measure the length of different objects and can help understand many mathematical concepts.

(a) Define the Sceneform fragment and listen for plane detection.
(b) After the plane surface is detected, the user can put two objects as AR Objects on the surface.
(c) Get the coordinates of the two objects and calculate the distance between those coordinates and show them using the text.

Summarizing the various features of our models as

(a) Using the local network easily controls desktop devices like keyboards and mice.
(b) Support for different platforms like android and windows with user friendly and easy to use user interface.
(c) Using OpenCV for colour detection makes it easy to detect and control the game.
(d) Using AR Core to detect the surface and place 3D models and add the interaction with 3D models
(e) Measure the length of the real-life object.

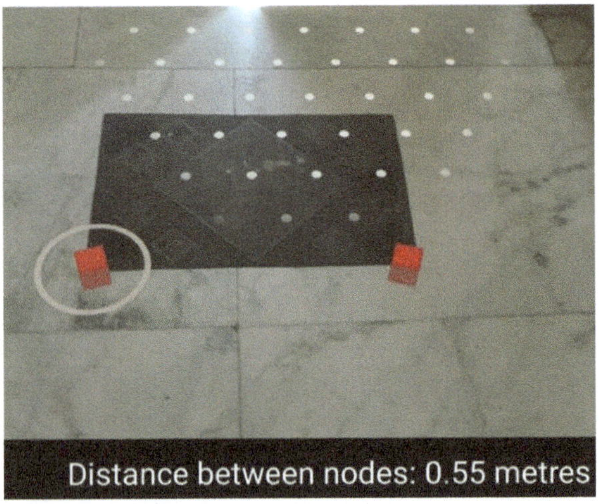

Fig. 7 Measure distance between components

We have used the concept of multithreading in our model for its functionality as compared to Android Remote Controller (ARC) [10, 14, 15] which hasn't incorporated this concept. Apart from that, we have also tried to provide an option for a handy feature like shut down and sleep the computer vision using the android application to make user accessibility over their computer. These features are not used in the ARC project and, we are using the computer vision for improving the gaming experience for the user which is not built by the ARC project. Some other features like for controlling the projector we have provided the volume key to change the slides of the presentation and many more. In the computer vision application, we are using a recently added tracking and colour detection algorithm which is contemporary with the other models. Our work includes a very user-friendly interface which helps the user to access all the features easily. In our work, we include the gaming in the android application either with gyro sensor or also providing the buttons for controlling the games through phones.

5 Conclusion and Future Work

We have proposed a model where multi-dimensional sub-systems are provided to the user. The model has its applicability in various domains from education, corporate, manufacturing industry etc. where the users can explore the system efficiently through an immersive environment. This environment is user-friendly and provides a very easily understandable view of concepts. It can help in reducing electronic waste by utilizing the bare minimum hardware like smartphones required to perform routine tasks. It also offers an option to the students to learn and understand the concepts more deeply using the latest advancements in technology. This is not only helping students to learn the concepts easily but also provide them with a platform to explore new dimensions for their learning. For the gamers, it provides a virtual immersive environment to interact through natural, body motions or gestures to provide a real-life experience.

In future work, the more advanced features of OpenGL and OpenCV may be explored through which better graphic options may be offered for the game. The latest tools are developed using Artificial Intelligence and machine learning which can be added to enhance the potential of learning to the next level. The application may be tested for various manufacturing units to appraise the interns in a better way before putting them to the real set of machinery.

References

1. Aarseth, E.: Computer game studies, year one. Game Stud. **1**(1), 1–15 (2001)
2. Abdinejad, M., Talaie, B., Qorbani, H.S., Dalili, S.: Student perceptions using augmented reality and 3D visualization technologies in chemistry education. J. Sci. Educ. Technol. **30**(1),

87–96 (2021)
3. Ahmed, K.T., Ummesafi, S., Iqbal, A.: Content based image retrieval using image features information fusion. Inf. Fusion **51**, 76–99 (2019)
4. Allen, J.G., Xu, R.Y.D., Jin, J.S.: Object tracking using Camshift algorithm and multiple quantized feature spaces. In: ACM International Conference Proceeding Series, vol. 100, pp. 3–7 (2004)
5. Alvarez-Santos, V., Iglesias, R., Pardo, X.M., Regueiro, C.V., Canedo-Rodriguez, A.: Gesture-based interaction with voice feedback for a tour-guide robot. J. Vis. Commun. Image Represent. **25**(2), 499–509 (2014)
6. Arslan, R., Kofoğlu, M., Dargut, C.: Development of augmented reality application for biology education. J. Turk. Sci. Edu. **17**(1), 62–72 (2020)
7. Azuma, R.T.: A survey of augmented reality. Pres. Teleoper. Virt. Environ. **6**(4), 355–385 (1997)
8. Bai, Z., Blackwell, A.F., Coulouris, G.: Through the looking glass: pretend play for children with autism. In: 2013 IEEE International Symposium on Mixed and Augmented Reality (ISMAR), pp. 49–58. IEEE (2013)
9. Banerjee, A., Ghosh, A., Bharadwaj, K., Saikia, H.: Mouse control using a web camera based on colour detection. arXiv:1403.4722
10. Barde, L., Dhole, N., Waghmare, P., Suryawanshi, S.: Controlling PC/laptop via android phone (android remote control). Int. J. Eng. Res. Technol. (IJERT) **3**(1).
11. Bertalmio, M., Sapiro, G., Randall, G.: Morphing active contours. IEEE Trans. Pattern Anal. Mach. Intell. **22**(7), 733–737 (2000)
12. Bradski, G., Kaehler, A.: Learning OpenCV: Computer Vision with the OpenCV Library. O'Reilly Media, Inc. (2008)
13. Craig, A.B., Sherman, W.R., Will, J.D.: Developing Virtual Reality Applications: Foundations of Effective Design. Morgan Kaufmann (2009)
14. Dahifale, S., Pandey, R., Ballani, R., Ingle, S.: Android desktop control (ADC). Int. J. Scient. Technol. Res. **3**(4) (2014)
15. Deshmukh, M., Jawale, D., Joshi, S., Kulkarni, P.S.: Android based wireless PC controller. Int. J. Comput. Sci. Inf. Technol. **6**(1) (2015)
16. Druzhkov, P.N., Erukhimov, V.L., Yu Zolotykh, N., Kozinov, E.A., Kustikova, V.D., Meerov, I.B., Polovinkin, A.N.: New object detection features in the OpenCV library. Pattern Recognit. Image Anal. **21**(3), 384 (2011)
17. DüNser, A., Billinghurst, M., Wen, J., Lehtinen, V., Nurminen, A.: Exploring the use of handheld AR for outdoor navigation. Comput. Graph. **36**(8), 1084–1095 (2012)
18. Escrivá, D.M., Joshi, P., Mendonça, V.G., Shilkrot, R.: Building Computer Vision Projects with OpenCV 4 and C++: Implement Complex Computer Vision Algorithms and Explore Deep Learning and Face Detection. Packt Publishing Ltd. (2019)
19. Fonseca, D., Redondo, E., Villagrasa, S.: Mixed-methods research: a new approach to evaluating the motivation and satisfaction of university students using advanced visual technologies. Univ. Access Inf. Soc. **14**(3), 311–332 (2015)
20. Gabbard, J.L., Swan, J.E., Hix, D., Kim, S.-J., Fitch, G.: Active text drawing styles for outdoor augmented reality: a user-based study and design implications. In: 2007 IEEE Virtual Reality Conference, pp. 35–42. IEEE (2007)
21. García, G.B., Suarez, O.D., Aranda, J.L.E., Tercero, J.S., Gracia, I.S., Enano, N.V.: Learning Image Processing with OpenCV. Packt Publishing Ltd. (2015)
22. Glover, J.: Unity 2018 Augmented Reality Projects: Build Four Immersive and Fun AR Applications Using ARKit, ARCore, and Vuforia. Packt Publishing Ltd. (2018)
23. Gollapudi, S.: Deep learning for computer vision. In: Learn Computer Vision Using OpenCV, pp. 51–69. Springer (2019)
24. Henrysson, A., Billinghurst, M., Ollila, M.: Face to face collaborative AR on mobile phones. In: Fourth Ieee and Acm International Symposium on Mixed and Augmented Reality (Ismar'05), pp. 80–89. IEEE (2005)

25. Ibáñez, M.B., Portillo, A.U., Cabada, R.Z., Barrón, M.L.: Impact of augmented reality technology on academic achievement and motivation of students from public and private Mexican schools. A case study in a middle-school geometry course. Comput. Edu. **145**, 103734 (2020)
26. Kalita, L.: Socket programming. Int. J. Comput. Sci. Inf. Technol. **5**(3), 4802–4807 (2014)
27. Kumar, V., Niyazi, K., Mahe, S., Vyawahare, S.: Mouse simulation using two coloured tapes (2012) arXiv:1204.1277
28. Laganière, R.: OpenCV Computer Vision Application Programming Cookbook, 2nd edn. Packt Publishing Ltd. (2014)
29. Lanham, M.: Learn ARCore-Fundamentals of Google ARCore: Learn to Build Augmented Reality Apps for Android, Unity, and the Web with Google ARCore 1.0. Packt Publishing Ltd. (2018)
30. Lee, D., Lee, S.G.: Vision-based finger action recognition by angle detection and contour analysis. ETRI J. **33**(3), 415–422 (2011)
31. Lee, S., Seo, Y.-H.: Vision-based virtual joystick interface. In: International Conference on Computational Science/Intelligence & Applied Informatics, pp. 65–78. Springer (2019)
32. Linowes, J., Babilinski, K.: Augmented Reality for Developers: Build Practical Augmented Reality Applications with Unity, ARCore, ARKit, and Vuforia. Packt Publishing Ltd. (2017)
33. López-Faican, L., Jaen, J.: Emofindar: evaluation of a mobile multiplayer augmented reality game for primary school children. Comput. Edu. **149**, 103814 (2020)
34. Malfatti, S.M., Dos Santos, F.F., Dos Santos, S.R.: Using mobile phones to control desktop multiplayer games. In: 2010 Brazilian Symposium on Games and Digital Entertainment, pp. 230–238. IEEE (2010)
35. Marco, L., Farinella, G.M.: Computer Vision for Assistive Healthcare. Academic Press (2018)
36. Morgan, A., Jones, Z., Chapman, R., Biaz, S.: Computer vision see and avoid simulation using OpenGL and OpenCV. Technical report CSS17–01. Auburn University
37. Murase, T., Moteki, A., Ozawa, N., Hara, N., Nakai, T., Fujimoto, K.: Gesture keyboard requiring only one camera. In: Proceedings of the 24th Annual ACM Symposium Adjunct on User Interface Software and Technology, pp. 9–10 (2011)
38. Pendleton, B.: Game programming with the Sdl. Linux J. **2003**(110), 1 (2003)
39. Sahin, D., Yilmaz, R.M.: The effect of augmented reality technology on middle school students' achievements and attitudes towards science education. Comput. Edu. **144**, 103710 (2020)
40. Saibaba, C.H.M.H., Waris, S.F., Raju, S.H., Sarma, V.S.R.K., Jadala, V.C., Prasad, C.: Intelligent voice assistant by using OpenCV approach. In: 2021 Second International Conference on Electronics and Sustainable Communication Systems (ICESC), pp. 1586–1593. IEEE (2021)
41. Shah, P., Pandya, K., Shah, H., Gandhi, J.: Survey on vision based hand gesture recognition (2019)
42. Shaikh, S., Thomas, K., Zuhair, S.: An exploratory study of E-waste creation and disposal: upstream considerations. Res. Conservat. Recycl. **155**, 104662 (2020)
43. Sherimon, P.C., Vinu, P.V., Krishnan, R.: Enhancing the learning experience in blended learning systems: a semantic approach. In Proceedings of the 2011 International Conference on Communication, Computing & Security, pp. 449–452 (2011)
44. Siegle, D.: Seeing is believing: using virtual and augmented reality to enhance student learning. Gift. Child Today **42**(1), 46–52 (2019)
45. Sin, A.K., Zaman, H.B.: Live Solar System (LSS): evaluation of an augmented reality book-based educational tool. In 2010 International Symposium on Information Technology, vol. 1, pp. 1–6. IEEE (2010)
46. Singh, S.K., Chauhan, D.S., Vatsa, M., Singh, R.: A robust skin color based face detection algorithm. Tamkang J. Sci. Eng. **6**(4), 227–234 (2003)
47. Souza Almeida, I. de, Oikawa, M.A., Carres, J.P., Miyazaki, J., Kato, H., Billinghurst, M.: AR-based video-mediated communication: a social presence enhancing experience. In: 2012 14th Symposium on Virtual and Augmented Reality, pp. 125–130. IEEE (2012)
48. Veluchamy, S., Karlmarx, L.R., Sudha, J.J.: Vision based gesturally controllable human computer interaction system. In: 2015 International Conference on Smart Technologies and Management for Computing, Communication, Controls, Energy and Materials (ICSTM), pp. 8–15. IEEE (2015)

49. Woo, M., Neider, J., Davis, T., Shreiner, D.: OpenGL Programming Guide: The Official Guide to Learning OpenGL, Version 1.2. Addison-Wesley Longman Publishing Co., Inc. (1999)
50. Yin, F., Makris, D., Velastin, S.A.: Performance evaluation of object tracking algorithms. In: IEEE International Workshop on Performance Evaluation of Tracking and Surveillance, Rio De Janeiro, Brazil, p. 25. Citeseer (2007)
51. Zhang, Z., Wu, Y., Shan, Y., Shafer, S.: Visual panel: virtual mouse, keyboard and 3D controller with an ordinary piece of paper. In: Proceedings of the 2001 Workshop on Perceptive User Interfaces, pp. 1–8 (2001)

Hand Gesture Recognition for Real-Time Game Play Using Background Elimination and Deep Convolution Neural Network

Kirti Aggarwal and Anuja Arora

Abstract In recent scenarios, Gaming Technology is on the surge of the utmost extent of technical advancement. Contemporary gaming using game controllers limits its reach due to various economic issues. The game programmers are making efforts towards developing games without a game controller. The expensive game controllers restrict the usage of gaming among users. In this research work, a mobile-based human–computer interaction system has been developed which enables users to control video games with more physical interaction and without more restrictions. The novelty of work is to play games using any good quality mobile/web camera instead of the game controller. A Convolution Neural Network (CNN) is combined with background elimination to detect different hand gestures and further uses these gestures to control video games. In the first instance, a background elimination algorithm is used to extract the hand gesture image captured through a mobile/web camera. These hand images have been used to train as well as predict the type of gesture. CNN is used to detect gestures and to render the appropriate motion in the game. The designed human–computer interaction system concludes 98.2% accuracy of hand gesture recognition for considered hand gestures. The measured latency is also adequate while game play.

Keywords Gesture recognition · Convolution neural network · Gaming control · Human–computer interaction · Computer vision

K. Aggarwal · A. Arora (✉)
Department of Computer Science and Engineering and Information Technology, Jaypee Institute of Information Technology, Noida, India

K. Aggarwal
e-mail: aggarwalkirti25@gmail.com

A. E. Hassanien et al. (eds.), *Virtual and Augmented Reality for Automobile Industry: Innovation Vision and Applications*, Studies in Systems, Decision and Control 412, https://doi.org/10.1007/978-3-030-94102-4_8

1 Introduction

Recent scenario demands Artificial Intelligence to overpower humans in almost all aspects such as e-learning, e-commerce, robotics, sustainability, etc. One similar kind of direction is game design and development. From the past few years, AI like OpenAI Five and human have started to compete against each other in many online multiplayer video games like DOTA-2 (Fig. 1a) but most of the time AI gets an unsuccessful ending in the struggle. Researchers and engineers of Google have adopted AI to allow humans to throw virtual punches and kicks in games like Mortal

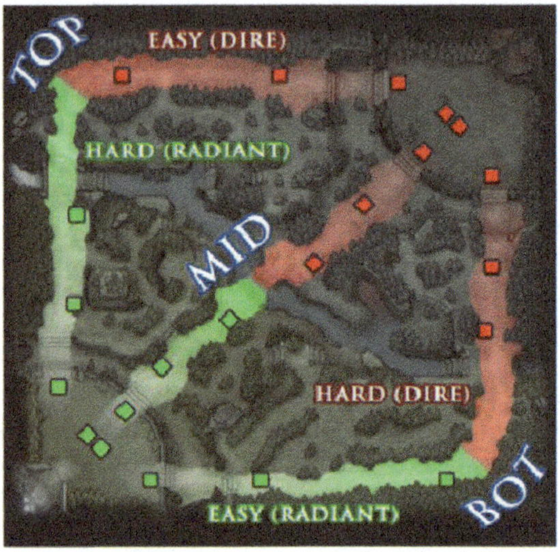

(a) The DotA 2 map [Source: https://dota2.fandom.com/wiki/Lane]

(b) Mortal Kombat 3 [Source: http://www.fightersgeneration.com/games/mk3.html]

Fig. 1 Screen shots of online video games

Kombat 3[1] (Fig. 1b). In this kind of game, instead of jabbing at the game controller, the user fights via a web camera. AI renders their airstrikes to an onscreen clone in real-time. It has become a common perception that playing games on mobile makes you couch potatoes and leads to obesity etc. But what if playing a video game becomes more of physical activity using the mobile device itself. Solutions to throw virtual punches in a game or playing tennis have already been implemented by companies like Microsoft Kinect. These Kinect Based systems are hardware-dependent, platform-dependent, and quite expensive solutions. Indeed, this will be a great insight that there are ways that enable us to control these video games with more physical interaction and without all the above hardware restrictions.

1.1 Research Motivation

Our work is an act in this direction to provide a less constrained interactive human–computer interaction that overcomes the extra hardware requirement and brings down the product cost [1]. The proposed work helps in enhancing the game controller system accessibility and reach to the customers. For such games, the hand is considered to be the best input device and powerful mode of communication to capture the gesture of human planned action. An existing game controller such as Play Stations or Kinect can easily be replaced by any high-quality mobile camera or web camera. Although numerous latest research works are carried out to recognize hand gestures, this area is not fully explored yet.

In our work, the specific hand gesture recognition problem dwells under the classification category i.e., the hand gesture recognition problem is tackled as a classification problem. 2D hand recognition has already been handled using machine learning and statistical modeling techniques such as Hidden Markov Model (HMM) [2, 3] and the Support Vector Machine [4]. One more well-known approach with effective hand gesture recognition is orientation histogram [5]. The orientation histogram algorithm is simple and fast to compare between feature vectors of gestures. This research work is a trial to capture and recognize real-time hand gestures and relevant action should be performed on the game screen.

1.2 Research Work Direction and Contribution

The proposed Human Game Interaction system is partitioned into three sections—First and foremost is 'Hand Recognition' which leads towards the second section 'Hand Gesture Identification' and the last and an important section is to operate according to recognized action. Various algorithms and techniques have been used to achieve overall effective outcomes such as background elimination technique is

---

[1] https://mortalkombat.fandom.com/wiki/Mortal_Kombat_3.

applied for hand recognition, Masking is used to generate hand gesture masks, deep learning models are used to classify hand gesture. All the processes and all their fragments are discussed in detail in the research paper. Even, Deep learning models along with computer vision techniques are used to build an unconstraint human game interaction system, and the identified gestures are used to render the appropriate motion in the games.

The prime aim is to measure the success of this system to achieve high accuracy in hand gesture recognition in real-time along with gameplay at a low cost. A reliable system should be able to control the game efficiently according to the player's hand gesture. The research contributions of this work are as follows:

- A deep learning-based hand gesture recognition system is designed which can recognize highly accurate hand gestures.
- The proposed system performs relevant game actions according to recognized hand gestures with the help of a mobile camera or web camera with low latency.

1.3 Paper Organization

The paper is structured as follows—Sect. 2 discusses related research work done in this direction. Next Sect. 3 lists the preliminaries and dataset used to perform these experiments. The proposed hand gesture recognition solution and work process flow are detailed in Sect. 4. Further, Sect. 5 discusses technical details used to make it a successful approach and discusses how background elimination and deep learning can make it an effective solution. Section 6 presents the experimental setup and results and finally, Sect. 7 is about the conclusion and future scope.

2 Related Work

This section in separated in two parts—first subsection discusses about recent research work done in automobile industry and second part is technical literature part which is relevant to work done in the presented chapter. Second subsection details about technology used while developing a automobile Game play to provide exposure of new drivers with the help of hand actions.

2.1 Augmented Reality and Virtual Reality for Game Play

Undoubtedly, Augmented Reality (AR) and Virtual Reality (VR) bridged the gap in digital and real world. These devices present the world visually in an adroit way flawlessly and user not able to differentiate in virtual and real world. Augmented reality is to bring digital concepts to live view or virtual settings whereas virtual reality is completely virtual world i.e. elements which are not visible in reality. Augmented reality is supplement of reality, it can provide feel of reality but can't completely replace it. Both augmented and virtual widen their spectrum in a number of application areas such as—Design, Manufacturing [6], Automobile [7], Construction [8], Entertainment [9], Education [10], sports [11], etc. Augmented reality is both interactive and present 3D environment. Basically, it combines real and virtual object to make application closer to real world.

Carmigniani et al. surveyed and presented state-of-art of AR technology, system, and applications. A number of computer vision methods, AR devices, AR interfaces, AR systems along with augmented reality mobile systems are discussed in detail [12]. For next generation mobile platforms, AR and VR devices recent advances and challenges towards optical and display specially for near-eye devices are discussed by Zhan et al. in 2020 [13]. One well-known and advanced application area in AR and VR is augmented reality. These advances in digital and information technologies transformed and arouse several open research challenges in environment research due to AR and VR technologies [14].

Augment and virtual reality provides support to wide variety of application areas as discussed above, but these technologies significantly altered the automobile industry. A large range of prototypes are developed for driver assistance to assist in location based service, driverless car, cruise control, obstacles identification [15], Driver behaviour [15], maintenance assistance etc. These AR embedded autonomous vehicles are helpful due to number of reasons, to help elderly and handicapped people for safe and confident drive [16, 17], help in managing incoming information (phone, car indications) receiving through varying channels [15], location tracking, decision making process etc. Due to all these open research problems and challenges automobile AR considered as a promising area and having high stakes. The autonomous cars contain sensors for various purposes such as safety, reminder, facility, assistance, etc. Where to present information gathered by sensors needs to be well driven in front of user so that he should take these as services not burden. These services should alleviate users' cognitive load and incorporate enjoyment, happiness, and comfort in driving [18]. Recent research direction for Augmented reality in automobile industry is growing towards- exploration of AR-HUD for convenient and safety in driving [19], Augmented Reality Simulation of Automotive Engine Assembly for Blended Learning [20], Education assistive technology in automobile industry to create and enrich experience by creating augmented reality of each and every part of car and to demonstrate working of engine [21], exhibit knowledge of driver response using augmented reality during an unexpected and unsafe scenario [22]. One open research issue is to exhibit car components behavior display through augmented reality in various situations. Interaction design through augmented reality in human machine interface of automobile industry is also underexplored area which needs researchers' attention [23]. Hence, automobile sector still needs exploration by researchers and still lots of augmented reality platforms are requisite to enrich automobile stakeholders' experience.

2.2 Hand Gesture Recognition Literature

In day-to-day life, hand gestures are a natural and powerful way of communication among humans especially people who have some difficulties in speaking or hearing or in scenarios where noise-less communication is preferred such as during a military operation army personnel may not want to make any noise because it can alert the enemy or during a medical operation a surgeon may assist nurse using hand gestures. Hand gestures also have the potential to be used in communication between humans and computers as they can have many applications like playing games or controlling a robot etc. Vision-based hand gesture recognition can have some advantages over traditional human–computer interaction (HCI) devices but recognition of shape and movement of the hands in images or videos is quite complex. Some related research papers and their complexity is summarized in this section.

In 2012 a hand gesture recognition system is introduced that uses Microsoft Kinect for Xbox [24]. The system uses a three-step process to recognize the gesture. In the first step for hand detection K-Mean Clustering, Graham Scan Algorithm, a modified version of Moore-Neighbor Tracing algorithm is used. In the second step finger Identification is done with the help of the Contour tracing algorithm and three-point alignment algorithm. In the last step, the Hand gesture is recognized using three layers of classifiers. Then Molchanov et al. have introduced a novel multi-sensor hand recognition system for an accurate and efficient car driver [25]. A color camera, a time of flight (TOF) depth camera, and short-range radar, and a Deep Neural Network classifier that takes input from the sensors are calibrated in the system. A true gesture occurs when the radar detects significant movement, radar in the system is always ON while the color and the depth camera are ON only for the gesture duration. Since the radar uses very low power, the overall energy requirement of the system is very less.

A real-time hand gesture-based interface for a mobile robot is introduced by Nagi et al. [26]. Max-Pooling Convolution Neural Network- MPCNN architecture using alternating convolution and max-pooling layers is used for supervised feature learning and classification of hand gestures given by humans to mobile robots using colored gloves. The MPCNN architecture is trained using online gradient descent to classify 6 gestures and it achieves 96% accuracy. Naguri and Bunescu introduced a gesture recognition system that is based on long short-term memory (LSTM) nd convolution neural network (CNN) [27].

Infrared sensors acquire 3D hand gestures and are input into the trained system. Leap Motion Controller is used to sense hand movements and then the gesture is detected using Recurrent Neural Network (RNN) and LSTM. Then the segmented gesture is classified into one of the predefined gestures using the Softmax model. A robust marker-less hand gesture recognition system is designed and the proposed system is used to track static as well as dynamic hand gestures [28]. Input to the system is captured using a webcam then some Data Image Processing (DIP) techniques for noise removal and smoothing of the image are used. After that contour extraction is done to detect the hand gesture. Gestures are recognized using convexity defects

Table 1 Deep learning based hand gesture recognition literature

Brief research details	References
Dataset: Thomas Moeslund's gesture recognition dataset Methodology: CNN and stacked denoising autoencoder	[29]
Dataset: self generated acquired via surface electromyography (sEMG) Methodology: convolution neural network (CNN) and recurrent neural network (RNN)	[30]
Interactive software game for hand and wrist exercise for software engineers Dataset: self collected 12,000 bare hand gesture images against a static dark background Methodology: convolution neural network	[31]
HandReha: game-based system for wrist rehabilitation Dataset: self generated dataset Methodology: hand region segmentation from static background then segmented region input to CNN model	[32]
Sign language gesture recognition with efficient hand gesture recognition Dataset: King Saud University Saudi Sign Language (KSU-SSL) dataset Methodology: data normalization, feature learning, feature fusion, multilayer perceptron, auto encoder	[33]
Fingertip detection and hand gesture recognition in real time RGB-D camera and 3D CNN Hand region extraction used in-depth skeleton-joint information images from a Microsoft Kinect Sensor version 2. Border tracing algorithm used to extract contours of the hands	[34]

then the detected hand gesture is mapped to their associated action. For dynamic gestures detection, the webcam detects palm for 5 continuous frames.

The latest work done in this direction using deep learning approaches is tabulated in Table 1.

3 Preliminaries and Dataset

Datasets for detecting gestures are not widely available and those which are available are very restricted to the features like skin color, radial signature [35], centroid distance [35], convexity defects, or curvature [36, 37], etc. All these techniques were needed in case of solving this problem using computer vision but deep learning has revolutionized computer vision in general. Deep learning is considered as a panacea to handle this sort of computer vision problem but needs lots and lots of data to train the system. Therefore, in this research work, we first describe the creation of our hand gesture library to learn out hand gesture features. In general, 16 hand gestures are considered in the gesture library for game play which gives freedom to perform various actions in a particular game [38].

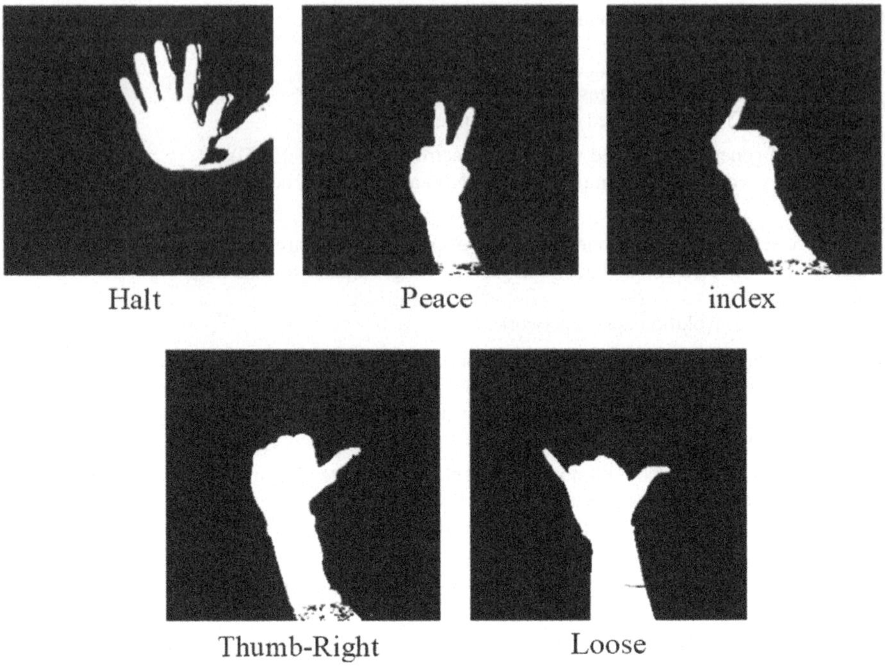

Fig. 2 Hand gesture dataset collected for training and test dataset

In our work, we have not worked on all 16 gestures to reduce the complexity of work and to get to know the effectiveness of deep learning models for hand gesture-based human game interaction systems. Our considered dataset has been generated for five defined hand gestures as shown in Fig. 2.

In total, approximately 8000 sample images are collected from 5 people. The number of training and test samples for all five hand gestures is mentioned in Table 1. One more dataset has been taken named as blank which is of all precipitous features which contain 826 training and 253 as test images.

Further, various data augmentation operations techniques are explored to extend the dataset, and a data level approach is used to generate an extra synthesized dataset through a sampling mechanism. Specifically, Python Augmenter Library [39] is used to apply Image processing operations such as skew, crop, rotate to obtain an extended dataset which is an extension of the dataset detailed in Table 2.

Table 2 Dataset details

Hand gesture	Training samples	Test samples
Halt	1626	313
Peace	16,261	312
Index	16,261	313
Thumb-right	16,261	313
Loose	783	157
Blank	826	253

4 Proposed Solution

The effectiveness of the targeted problem depends not only on the robustness of hand gesture recognition, but reduced latency time of hand gesture correspondent game play action is a requisite factor. Undoubtedly, reduced latency depends on various factors—dataset, pre-processed techniques applied to retain dataset, gesture recognition methodology, and streaming with the game without adversely affecting system reliability.

The architecture of the proposed hand gesture recognition is shown in Fig. 3. As shown in the figure, the initial step in hand recognition is to eliminate the undesirable area and determine the hand region. A process for background elimination has been performed for the same. The outcome of the process is foreground mask formation which will be the input for the deep learning model to identify/recognize the action performed in the specified hand mask. Further, the game must execute activity inside the gaming screen for a particularly defined hand action with minimal latency as game play is considered a real-time application.

5 Methodology

5.1 Background Elimination

Background subtraction is a mainstream algorithm for moving object detection in video surveillance systems. It segments moving objects by using the difference between the background and input images. The key to background subtraction is to fetch a reliable initial background as shown in Fig. 4. In this study, we have used a background subtraction algorithm based on category entropy analysis that dynamically creates color categories for each pixel in the images.

The category entropy analysis algorithm uses the concept of a joint category to build background categories that can adapt to the color disturbance of the background. The background is masked by moving objects in the background extraction phase itself. The category entropy analysis algorithm calculates color difference with color category. If the distance of color on two pixels is more than the threshold category,

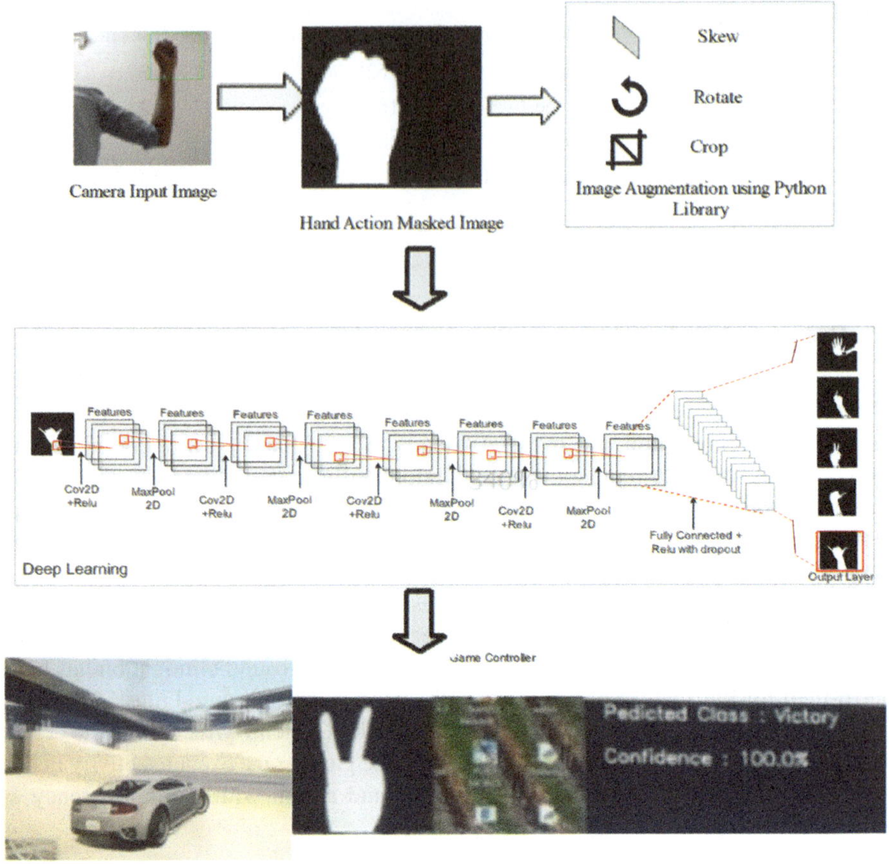

Fig. 3 Process flow of hand gesture recognition for game controller

then add this to a new color category otherwise classify it as a joint category. Compute joint category entropy to select candidate background category. In the proposed work, OpenCV has been used to calculate the average of the background for 30 frames. Further, the computed average has been used to detect the hand that has to be introduced after the background has been properly recognized.

5.2 Deep Convolution Neural Network

The next step is to recognize action performed by hand and design a well-defined and explanatory convolution neural network for the hand gesture recognition problem. A convolution neural net is used to extracts features from 2D data and assign weights to those features, eventually resulting in a prediction. CNN is a specific type of

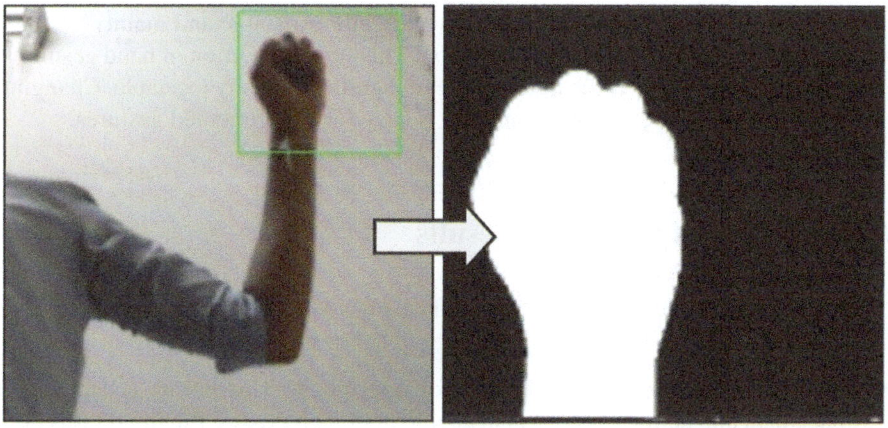

Fig. 4 Background elimination

artificial neural network that uses perceptron—a machine learning unit algorithm for supervised learning and TensorFlow—an open-source software library that is used for dataflow programming across a range of tasks. TensorFlow is a symbolic math library that is used for deep learning applications such as neural networks [40].

The proposed CNN network contains seven hidden convolution layers with Relu as the activation function and one fully connected layer. The network is trained across 50 iterations with a batch size of 64. Max pooling operation is applied in each layer which prevents the model from overfitting. Even, max pooling is important to predict gestures for the images that have not been used earlier. Further, Adam optimizer is used instead of the classical stochastic gradient descent procedure to update network weights iterative based on training data. We noticed that 50 iterations train the model well and there is no increase in validation accuracy along the lines so that should be enough. Some pre-processing operations have also been applied to embed contrast, scale, and transition to background eliminated hand gesture images. Threshold, resize and center of hand image are the applied preprocessing steps and these steps helped to train the introduced CNN model effectively.

5.3 Human–Computer Interaction

Games such as GTA, Road Redemption, FIFA, Mortal Kombat can be interactive and cost-effective by using the proposed experiments. No external interface/support is required. Some forethoughts are attention-seeking while experimentation. The prime concern is to control the mouse cursor by hand gesture. For the same and to resolve this same problem, we have selected specific five hand gestures. While performing the action or playing the game, the hand will be almost stable and mainly gestures are either of palm or different fingers' action. The topmost portion of hand gesture is our point of interest as specific figure shape recognition is easily locatable. Changing hand gestures from one gesture to another is easy for experimented gestures.

6 Experiments Setup and Results

For deep Learning, the Keras library is used which is a high-level API of TensorFlow. Some are as follows:

- ImageDataGenerator[2] library is used for image data. This library accepts input image data and applies various data augmentation processes. Finally returns both the initial image library and augmented image library.
- Deep Learning libraries such as Convolution2D, MaxPooling2D, Flatten, Dense, etc. are used to apply deep learning models for hand gesture recognition.
- Early Stopping of Callback from Keras has been used to stop the training when there is no improvement in the number of the validation loss.

Our proposed model can recognize hand gestures with 98.2% accuracy while playing the game. A screenshot of the hand gesture, its recognized class along with confidence and corresponding Game action is shown in Fig. 5a, b.

Figure 5a shows a particular hand gesture which is recognized with 99.81% confidence as a Palm hand gesture and shows car movement action corresponding to this same hand gesture i.e., straight movement of the car while game play. Another result is shown in Fig. 5b which is a victory hand gesture and recognized with 100% confidence and the car takes turn according to the applied hand gesture.

The proposed hand gesture system for game play is tested for five taken hand gestures on GTA and Road redemption game and the proposed system can perform game action without noticeable latency.

7 Conclusion and Future Scope

In this work, a real-time hand gesture recognition system for gameplay is implemented. The proposed work contains quite extensive tasks starting from data collection to gameplay. Initially, hand gesture data is collected for five specific hand gestures of different people. Further, the category entropy analysis algorithm for background elimination is done for hand gesture masking and the Convolution neural network deep learning model is applied for the recognition of masked hand images. In the end, recognized hand gesture corresponding game action is performed on the game screen.

[2] https://www.tensorflow.org/api_docs/python/tf/keras/preprocessing/image/ImageDataGenerator.

(a) Palm hand gesture

(b) Victory hand gesture

Fig. 5 Screen shot of hand gesture, predicted hand gesture along with prediction confidence and corresponding game play action

The system is validated for two-game applications where the achieved accuracy of gesture recognition within 10 epochs is 98.2% without observable latency/delay time.

This project can be transformed in the future for applications beyond gaming. Even less attuned video gaming areas may use the proposed system for various applications such as in schools, workplaces, or hospitals. Video games for training soldiers and teaching physics and camera drive control programs for physiotherapists for their patients' treatment can also be implemented in the future.

The proposed system could potentially work without a calibrating step and they should be able to detect several people in a wide area in front of the TV screen, so people can move around without worrying about leaving the "sight" of the controller. Users tend to play differently and can be automatically categorized into different player types based on their interaction with the game. In a large-scale study, data could be collected and used to automatically identify player types based on their gesture interaction behavior and allow dynamic and automatic settings and mappings to optimize setups. This would allow the generation and mapping of gesture-based user types to speed up learning of interaction with the device easier.

References

1. Kaur, S., Gupta, A., Aggarwal, A., Gupta, D., Khanna, A.: Sign language recognition using Microsoft Kinect. In: International Conference on Innovative Computing and Communications, pp. 637–645. Springer, Singapore (2021)
2. Yang, R., Sarkar, S.: Gesture recognition using hidden Markov models from fragmented observations. In: 2006 IEEE Computer Society Conference on Computer Vision and Pattern Recognition (CVPR'06), vol. 1, pp. 766–773. IEEE (2006)
3. Min, B.W., Yoon, H.S., Soh, J., Yang, Y.M., Ejima, T.: Hand gesture recognition using hidden Markov models. In: 1997 IEEE International Conference on Systems, Man, and Cybernetics. Computational Cybernetics and Simulation, vol. 5, pp. 4232–4235. IEEE (1997)
4. Neethu, P.S., Suguna, R., Sathish, D.: An efficient method for human hand gesture detection and recognition using deep learning convolutional neural networks. Soft Comput. 1–10 (2020)
5. Freeman, W.T., Roth, M.: Orientation histograms for hand gesture recognition. In: International Workshop on Automatic Face and Gesture Recognition, vol. 12, pp. 296–301. IEEE Computer Society, Washington, DC (1995)
6. Nee, A.Y., Ong, S.K., Chryssolouris, G., Mourtzis, D.: Augmented reality applications in design and manufacturing. CIRP Ann. **61**(2), 657–679 (2012)
7. Nag, B., De, D.: The Indian automobile industry: technology enablers preparing for the future. New Frontiers of the Automobile Industry, pp. 301–321. Palgrave Macmillan, Cham (2020)
8. Noghabaei, M., Heydarian, A., Balali, V., Han, K.: Trend analysis on adoption of virtual and augmented reality in the architecture, engineering, and construction industry. Data **5**(1), 26 (2020)
9. Hung, S.W., Chang, C.W., Ma, Y.C.: A new reality: exploring continuance intention to use mobile augmented reality for entertainment purposes. Technol. Soc. **67**, 101757 (2021)
10. Iatsyshyn, A.V., Kovach, V.O., Lyubchak, V.O., Zuban, Y.O., Piven, A.G., Sokolyuk, O.M., Shyshkina, M.P., et al.: Application of augmented reality technologies for education projects preparation (2020)
11. Soltani, P., Morice, A.H.: Augmented reality tools for sports education and training. Comput. Educ. **155**, 103923 (2020)
12. Carmigniani, J., Furht, B., Anisetti, M., Ceravolo, P., Damiani, E., Ivkovic, M.: Augmented reality technologies, systems and applications. Multimed. Tools Appl. **51**(1), 341–377 (2011)
13. Zhan, T., Yin, K., Xiong, J., He, Z., Wu, S.T.: Augmented reality and virtual reality displays: perspectives and challenges. Iscience 101397 (2020)
14. Trofimenko, Y.: Environmental problems of using digital and information technologies in automobile transport. In: MATEC Web of Conferences, vol. 341, p. 00069. EDP Sciences (2021)
15. George, P., Thouvenin, I., Fremont, V., Cherfaoui, V.: DAARIA: driver assistance by augmented reality for intelligent automobile. In: 2012 IEEE Intelligent Vehicles Symposium, pp. 1043–1048. IEEE (2012)

16. Rashid, Z., Melià-Seguí, J., Pous, R., Peig, E.: Using augmented reality and internet of things to improve accessibility of people with motor disabilities in the context of smart cities. Futur. Gener. Comput. Syst. **76**, 248–261 (2017)
17. Schall, M.C., Jr., Rusch, M.L., Lee, J.D., Dawson, J.D., Thomas, G., Aksan, N., Rizzo, M.: Augmented reality cues and elderly driver hazard perception. Hum. Factors **55**(3), 643–658 (2013)
18. Ng-Thow-Hing, V., Bark, K., Beckwith, L., Tran, C., Bhandari, R., Sridhar, S.: User-centered perspectives for automotive augmented reality. In: IEEE International Symposium on Mixed and Augmented Reality, pp. 13–22 (2013)
19. Zhang, M.: Optimization analysis of AR-HUD technology application in automobile industry. J. Phys. Conf. Ser. **1746**(1), 012062. IOP Publishing (2021)
20. Kang, M.S.: Development of automotive engine assembly augmented reality simulation for blended learning. J. Ind. Converg. **18**(1), 17–23 (2020)
21. Karaket, N., Auephanwiriyakul, S., Theera-Umpon, N.: Automobile parts localization using multi-layer multi-model images classifier ensemble. Advances in Intelligent Information Hiding and Multimedia Signal Processing, pp. 367–376. Springer, Singapore (2021)
22. Smith, M., Jordan, L., Bagalkotkar, K., Sai Manjuluri, S., Nittala, R., Gabbard, J.: Hit the brakes! Augmented reality head-up display impact on driver responses to unexpected events. In: 12th International Conference on Automotive User Interfaces and Interactive Vehicular Applications, pp. 46–49 (2020)
23. Zhang, C.: Application research of interaction design in human-machine interface of automobile. In: International Conference on Intelligent and Interactive Systems and Applications, pp. 405–412. Springer, Cham (2020)
24. Li, Y.: Hand gesture recognition using Kinect. In: 2012 IEEE International Conference on Computer Science and Automation Engineering, pp. 196–199. IEEE (2012)
25. Molchanov, P., Gupta, S., Kim, K., Pulli, K.: Multi-sensor system for driver's hand-gesture recognition. In: 2015 11th IEEE International Conference and Workshops on Automatic Face and Gesture Recognition (FG), vol. 1, pp. 1–8. IEEE
26. Nagi, J., Ducatelle, F., Di Caro, G.A., Cireşan, D., Meier, U., Giusti, A., Gambardella, L.M.: Max-pooling convolutional neural networks for vision-based hand gesture recognition. In: 2011 IEEE International Conference on Signal and Image Processing Applications (ICSIPA), pp. 342–347. IEEE (2011)
27. Naguri, C.R., Bunescu, R.C.: Recognition of dynamic hand gestures from 3D motion data using LSTM and CNN architectures. In: 2017 16th IEEE International Conference on Machine Learning and Applications (ICMLA), pp. 1130–1133. IEEE (2017)
28. Haria, A., Subramanian, A., Asokkumar, N., Poddar, S., Nayak, J.S.: Hand gesture recognition for human-computer interaction. Procedia Comput. Sci. **115**, 367–374 (2017)
29. Oyedotun, O.K., Khashman, A.: Deep learning in vision-based static hand gesture recognition. Neural Comput. Appl. **28**(12), 3941–3951 (2017)
30. Nasri, N., Orts-Escolano, S., Cazorla, M.: An SEMG-controlled 3D game for rehabilitation therapies: real-time time hand gesture recognition using deep learning techniques. Sensors **20**(22), 6451 (2020)
31. Rungruanganukul, M., Siriborvornratanakul, T.: Deep learning based gesture classification for hand physical therapy interactive program. In: International Conference on Human-Computer Interaction, pp. 349–358. Springer, Cham (2020)
32. Farahanipad, F., Nambiappan, H.R., Jaiswal, A., Kyrarini, M., Makedon, F.: HAND-REHA: dynamic hand gesture recognition for game-based wrist rehabilitation. In: Proceedings of the 13th ACM International Conference on PErvasive Technologies Related to Assistive Environments, pp. 1–9 (2020)
33. Al-Hammadi, M., Muhammad, G., Abdul, W., Alsulaiman, M., Bencherif, M.A., Alrayes, T.S., Mekhtiche, M.A., et al.: Deep learning-based approach for sign language gesture recognition with efficient hand gesture representation. IEEE Access **8**, 192527–192542 (2020)
34. Tran, D.S., Ho, N.H., Yang, H.J., Baek, E.T., Kim, S.H., Lee, G.: Real-time hand gesture spotting and recognition using RGB-D camera and 3D convolutional neural network. Appl. Sci. **10**(2), 722 (2020)

35. Trigueiros, P., Ribeiro, A.F., Reis, L.P.: A comparative study of different image features for hand gesture machine learning (2013)
36. Suzuki, S.: Topological structural analysis of digitized binary images by border following. Comput. Vis. Graph. Image Process. **30**(1), 32–46 (1985)
37. Sklansky, J.: Finding the convex hull of a simple polygon. Pattern Recogn. Lett. **1**(2), 79–83 (1982)
38. Xu, P.: A real-time hand gesture recognition and human-computer interaction system (2017). arXiv preprint. arXiv:1704.07296.
39. Bloice, M.D.: Augmentor (2016) [Online]. http://augmentor.readthedocs.io
40. Baranwal, S., Arora, A., Khandelwal, S.: Detecting diseases in plant leaves: an optimised deep-learning convolutional neural network approach. Int. J. Environ. Sustain. Dev. **20**(2), 166–188 (2021)

Modeling of Optimal Bidirectional LSTM Based Human Motion Recognition for Virtual Reality Environment

R. Thenmozhi, Shabnam Mohamed Aslam, A. Arokiaraj Jovith, and T. Avudaiappan

Abstract Recently, virtual reality (VR) has received significant attention among researchers to accomplish realistic interaction between humans and computers. Among the different application areas of VR, multimedia human computer interaction requires accurately collect and identify the motions of human body in real time. For enabling highly realistic and efficient communication among humans and computers, design of human motion recognition system becomes necessary to determine diverse complex and distinct human actions. Therefore, this study designs a new optimal bidirectional long short term memory (BiLSTM) with fully convolution network (FCN), called OBiLSTM-FCN model for human motion recognition in VR environment. The proposed OBiLSTM-FCN model comprises different processes namely feature extraction, classification, and hyperparameter optimization. Primarily, kernel based linear discriminant analysis (LDA) approach is employed as a feature extraction technique. In addition, the BiLSTM-FCN technique is applied as a recognition model to determine human motions. Finally, the Adam optimizer is applied to optimally tune the hyperparameters involved in the BiLSTM-FCN model. The performance validation of the OBiLSTM-FCN model take place and the resultant values portrayed the improved performance over the other compared methods.

R. Thenmozhi (✉)
Computing Technologies, SRM Institute of Science and Technology, Kattankulathur 603203, India
e-mail: thenmozr@srmist.edu.in

S. M. Aslam
Department of Information Technology, Majmaah University, Al Majmaah 11952, Saudi Arabia
e-mail: s.aslam@mu.edu.sa

A. A. Jovith
Department of Networking and Communications, SRM Institute of Science and Technology, Kattankulathur 603203, India
e-mail: arokiara@srmist.edu.in

T. Avudaiappan
Department of Computer Science and Engineering, K. Ramakrishnan College of Technology, Trichy 621112, India
e-mail: avudaiappant.cse@krct.ac.in

A. E. Hassanien et al. (eds.), *Virtual and Augmented Reality for Automobile Industry: Innovation Vision and Applications*, Studies in Systems, Decision and Control 412, https://doi.org/10.1007/978-3-030-94102-4_9

Keywords Virtual reality · Deep learning · Human motion recognition · Hyperparameter tuning · Adam optimizer

1 Introduction

In recent years, Virtual reality (VR), the utilization of computation technology to create a simulated environment, has spiked in use. To render the virtual world from the user perspective, the users' location needs to be tracked and calculated [1]. Each VR system measures head position, head orientation, and most of the measures hand position and orientation. Since consumer-level VR device becomes affordable, widely available, and more powerful, VR based experience discover additional applications in gaming, industry, and entertainment [2]. Currently, the VR device provides an incorporated higher-end tracing of head and two-handed movement off the shelf, without resorted to further movement capture gear. These 3 fundamental tracking points previously offer sufficient data regarding an action performing in a VR experience and user's motion [3]. Indeed, the incidental motion data traced at the time of VR session have significant attention for a multitude of VR-based movement analyses task that is applicable for many commercial, private, as well as scientific and medical applications. e.g., assessment and evaluation in virtual training sessions, optimization, and performance analysis in task-oriented VR scenarios. The certain key quality of this kind of application has a stronger connection amid complex and rich motion information and it was recorded in the context of virtual environments [4]. This possesses novel challenges for the study of human motion recorded in VR. Figure 1 depicts the framework of VR model.

Motion capture technique represents the usage of acoustic, optical, or another motion data gathering tool for recording the action of the performers and succeeding processing to transform the recorded information to data which could drive the movement of the computer virtual characters [5]. The data gathered by tracing could be

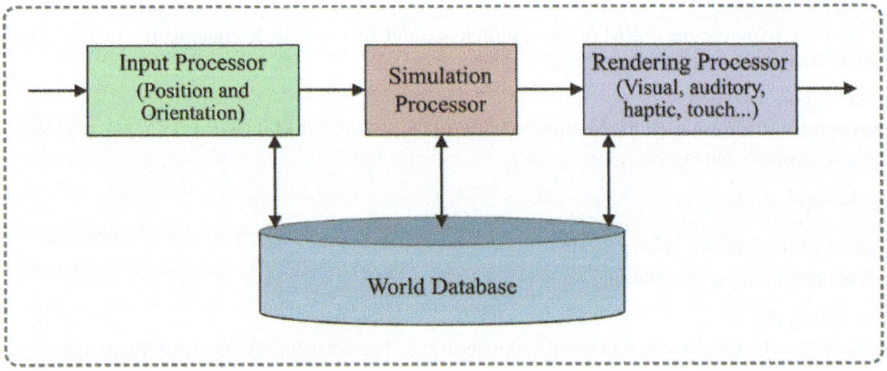

Fig. 1 Structure of virtual reality system

utilized for simply describing the location in certain 3D spaces of the captured object's limb, or it is utilized for describing the subtle variations in the complicated skin/face [6]. With the fast growth of computer technologies, the New York Computer Graphics Technology Laboratory has developed mercury mirror optical devices for projecting the efficiency of the real dancer on the computer display as an orientation for the main structure of digital dancer animations [7]. Also, it helps the growth of movement gesture capturing techniques. Movement gesture capturing techniques have rapidly gained the consideration of many developers and authors [8].

Zhang et al. [9] build a VR technique that employs a VR tool for gathering actions, and with the movement information of the VR technique, the human demonstration is inferred to an action series which is executed with the help of robots. Jiang et al. [10] present a new real-time motion rebuilding and detection technique in VR with the orientations and positions of 2 hands and the user's head. To rebuild natural motion based on rare sensors, they separate the entire body to the lower and upper bodies. While a natural action detection method based on NN would operate for detecting the target motions.

Fu et al. [11] presented an attitude evaluation method adopted to be implanted. The preceding centralized Kalman filter is alienated to 2-phase Kalman filters. Based on the distinct features of the sensor, they are separately treated for isolating the cross-influence among the sensor nodes. An adoptive adjustment technique-based FL method has been proposed. The geomagnetic field strength, angular velocity, and acceleration of the environments are utilized as the input of the FL method for judging the motion state and later alter the covariance matrices of the filter. The adaptive change of the sensors is transformed to the detection of the movement activity. Fazeli et al. [12] presents a study in the growth of virtual environments to analyse hand motions for evaluating patients' hand movement. With uncontacted motion sensor nodes in the enhanced virtual environments, information for the hand restoration is precisely attained. Leap Motion sensor and Unity 3D are utilized as an instrument for real-time information capturing of hand motions, which provide graphical data for angles, hand fingers position, gesture types and grab strength.

Kritikos et al. [13] aim is to estimate the sentiment of existence in distinct hardware setups of VR Exposure Therapy, as well, mainly how the users interact with this setup could affect their sense of incidence in the virtual experiment. The acrophobic virtual scenarios are utilized as an analysis via twenty phobic persons and the Witmer Singer presence questionnaires have been utilized for incidence assessment by the user. In Sivasamy et al. [14], a new continuous validation method is performed for the user in the VR environments. In this technique, head movement is utilized for providing continuous authentication. VRC Auth employs distinct ML approaches for classifying the head movement of unauthorized and authorized users.

In Liu et al. [15], a hybrid Gaussian method has been utilized for building a background method in real-time based variations in VR data, also the images are deducted with the background differential model. The optical flow technique has been utilized for foreground recognition of the targets, and the effect of dense and sparse optical flows are related to obtaining the movement features and optical flow data of the targeted body, correspondingly. Zhou [16] present an enhanced method of the 3-channel stable extremal regions to enhance the 3-channel stable extremal regions. The method could adoptively chooses the filter of all the channels for filtering the feature region extracted from the 3 stable extremal values. Generally, an action cycle is 30–50 frames, hence it is fast and beneficial to straightforwardly utilize the 1st fifty frames of video for processing.

This study designs a new optimal bidirectional long short term memory (BiLSTM) with fully convolution network (FCN), called OBiLSTM-FCN model for human motion recognition in VR environment. The proposed OBiLSTM-FCN model comprises different processes namely feature extraction, classification, and hyper-parameter optimization. The proposed model utilizes kernel based linear discriminant analysis (LDA) approach for feature extraction technique. Moreover, the BiLSTM-FCN technique is applied as a recognition model to determine human motions. Furthermore, the Adam optimizer is applied to optimally tune the hyper-parameters involved in the BiLSTM-FCN model. The performance validation of the OBiLSTM-FCN model takes place and the results are examined under different dimensions.

2 The Proposed Human Motion Recognition Model

This study has designed an effective OBiLSTM-FCN model for human motion recognition in VR environment. The proposed OBiLSTM-FCN model comprises different processes namely kernel LDA based feature extraction, BiLSTM-FCN based classification, and Adam hyperparameter optimization.

2.1 Kernel LDA Based Feature Extraction

LDA has been linear technique usually utilized to feature extraction. The LDA technique was insensitive for altering illumination and attitude and so extremely utilized from image recognition tasks. But, this technique as typical LDA is essentially linear. Because of the difficulty and diversity of human motion from VR scenes, any major high dimension non-linear feature data hidden from the motion information could not be removed. Thus, this work establishes a kernel function from the LDA technique to non-linear presented for extracting expression features. Afterward, the genetic optimization SVM classifier the difficult performance classification and identification has been lastly gathered.

During the human motion data removal applications, assume that A exists the active matrix. In the LDA technique, A refers to the complete rank matrix with class label [17]:

$$A = [a_1 \dots a_n] = [B_1 \dots B_k] \in R^{m \times n} \tag{1}$$

Amongst them, all $a_i (1 \leq i \leq n)$ has been data point from the m-dimension space. All the block matrix $B_i \in R^{m \times n} (1 \leq i \leq k)$ has been gathered of data items from the i th class. n_i implies the size of class i and the entire amount of data items from dataset A is n. Consider N_i represent the column index fitting to class i. The global center c of A and local center c_i of all the classes A_i are correspondingly written as:

$$c = \frac{1}{n} Ae, c_i = \frac{1}{n_i} B_i e_i, i = 1, \ldots, k \tag{2}$$

Let

$$S_b = \sum_{i=1}^{k} n_i (c_i - c)(c_i - c)^T \tag{3}$$

$$S_w = \sum_{i=1}^{k} \sum_{j \in N_i} (a_j - c)(a_j - c)^T \tag{4}$$

$$S_t = \sum_{j=1}^{n} (a_j - c)(a_j - c)^T \tag{5}$$

Amongst them, S_b, S_w and S_t demonstrated as inter-class divergence matrix, intra-class divergence matrix, and total divergence matrix correspondingly.

$$S_t = S_b + S_w \tag{6}$$

Afterward, the typical LDA objective function has appeared this:

$$G = \underset{G \in R^{m \times 1}}{\arg\max}\, trace\left(\left(G^T S_t G \right)^{-1} \left(G^T S_b G \right) \right) \tag{7}$$

It can appreciate that LDA technique was fundamentally a linear technique, therefore the outcome is not great if controlling non-linear issues, and there are singularities. For effectively extracting the non-linear features of the data, it can utilize the kernel decision LDA for extracting features.

2.2 *BiLSTM-FCN Based Classification*

The derived feature vectors are passed into the BiLSTM-FCN model to accomplish classification process. Hochreiter and Schmidhuber presented the LSTM mainly to conquer the gradient vanishing problems. LSTM has been variation of RNNs which contain similar kinds of output and input. But, apart from RNN, LSTM consists of output, input, and forget gates. Hence, it could manage what has to be forgotten and what has to be kept. For that reason, LSTM could keep data from the historical, where RNN can't. Generally, computations within the LSTM method produce in keeping with the succeeding calculation i.e., achieved at every time step. This calculation offers the comprehensive method for a current LSTM with forgetting gate:

$$g^{(z)} = \tanh\left(W^{gx}x^{(z)} + W^{gh}h^{(z-1)} + b_g\right) \tag{8}$$

$$i^{(z)} = \sigma\left(W^{ix}x^{(z)} + W^{ih}h^{(z-1)} + b_i\right) \tag{9}$$

$$f^{(z)} = \sigma\left(W^{fx}x^{(z)} + W^{fh}h^{(z-1)} + b_f\right) \tag{10}$$

$$o^{(z)} = \sigma\left(W^{ox}x^{(z)} + W^{oh}h^{(z-1)} + b_o\right) \tag{11}$$

$$s^{(z)} = g^{(z)} \odot i^{(i)} + s^{(z-1)} \odot f^{(z)} \tag{12}$$

$$h^{(z)} = \tanh\left(s^{(z)}\right) \odot o^{(z)} \tag{13}$$

whereas x(z) denotes the input layer at present time step z, h(z) denotes the values of hidden layer of LSTM, in which h(z − 1) indicates resultant value by every memory cell from the hidden layer at prior time [18]. σ signifies the sigmoid function, tanh, and $\odot$ means hyperbolic tangent function and element-wise multiplication, correspondingly. BiLSTM processes sequential data in all backward and forward directions with 2 distinct hidden layers for capturing future and past data, correspondingly; next, the 2 hidden states are concatenated to generate the last output. Figure 2 illustrates the framework of BiLSTM model.

The CNN is influential graphic model which has the capacity of yielding hierarchies of features [19]. FCN is largely used in the temporal domain and has terminated to be helpful to handle the temporal dimension for TSC without any feature engineering and immense data pre-processing. In order to univariate TSC, FCN can be defined by:

$$z = w \odot x + b \tag{14}$$

$$\mathrm{a} = \mathrm{BN(z)} \tag{15}$$

$$y = ReLU(a) \tag{16}$$

In which w, x, and b represents tensor, input vector, and bias vector at time-step t, correspondingly, and $\odot$ denotes the convolutional operators. The FCN framework comprises 3 convolutions of one dimensional kernel (8, 5, and 3). All the blocks are followed by a ReLU activation layer and BN. Afterward the convolution

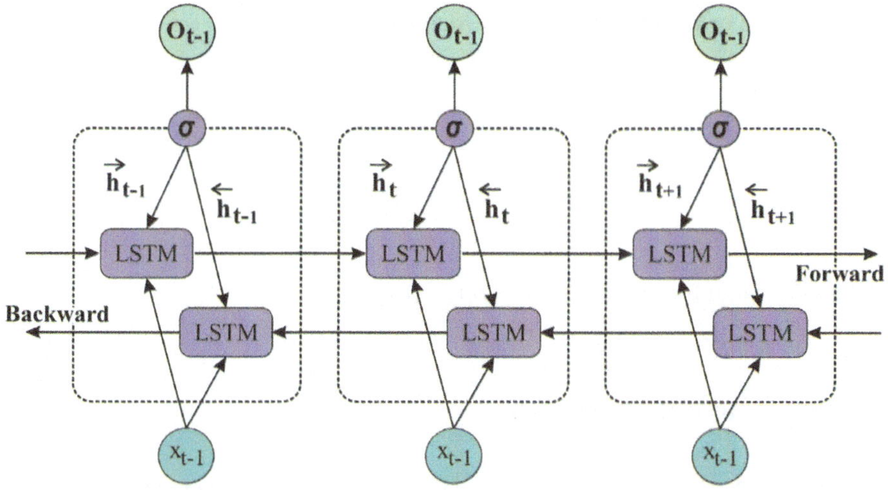

Fig. 2 Structure of BiLSTM

block, feature is fed to global average pooling layers. Next, the concluding label y is generated in the softmax layers.

They increase BiLSTM and FCM to suggest end-to-end hybrid methods because a BiLSTM consider backward and forward dependencies in time sequence data. The presented method comprises 2 branches, BiLSTM, and FCN. Initially, convolution method is utilized as a feature extractor; this block contains 3 convolutions one dimensional kernel with the sizes (8, 5, and 3) without striding. Afterward the convolutional block, the feature is feed to a global average pooling layer. The next branch contains BiLSTM blocks. BiLSTM train 2 LSTM models rather than one model that could access long-range context in backward and forward directions in time sequence data.

2.3 Adam Based Hyperparameter Optimization

For optimal selection of the hyperparameters involved in the BiLSTM-FCN model, the Adam optimizer is used. It is generally applied to determine the adaptive learning values in which the variables are employed to train the DL model. It is simple for the first order gradient with restricted memory. It is widely used based on the gradient descent and momentum approach. Therefore, the first order momentum can be accomplished using Eq. (17):

$$m_i = \beta_1 m_{i-1} + (1 - \beta_1)\frac{\partial C}{\partial w}. \tag{17}$$

The second order momentum can be denoted as follows,

$$v_i = \beta_2 v_{i-1} + (1 - \beta_2)\left(\frac{\partial C}{\partial w}\right)^2. \tag{18}$$

$$w_{i+1} = w_i - \eta \frac{\hat{m}_i}{\sqrt{\hat{v}_i + \epsilon}}, \tag{19}$$

where $\hat{m}_i = m_i/(1 - \beta_1)$ and $\hat{v}_i = v_i/(1 - \beta_2)$.

3 Experimental Validation

The performance validation of the OBiLSTM-FCN technique takes place on the classification of diverse human movements in VR environment. The results are inspected under ten different types of motions. Table 1 and Fig. 3 provide a comparative precision analysis of the OBiLSTM-FCN technique with other techniques. The results portrayed that the OBiLSTM-FCN technique has gained effective outcomes with higher precision values compared to other techniques. For instance, with motion type 2, the OBiLSTM-FCN technique has obtained an increased precision of 0.9943 whereas K-means-SVM and LDA-GA-SVM techniques have achieved a reduced precision of 0.9463 and 0.9913. Similarly, with motion type 5, the OBiLSTM-FCN technique has attained a maximum precision of 0.9869 whereas K-means-SVM and LDA-GA-SVM techniques have resulted in minimum precision of 0.9241 and

Table 1 Precision analysis of OBiLSTM-FCN model with varying motion type

Motion type	Precision		
	K-means-SVM	LDA-GA-SVM	OBiLSTM-FCN
1	0.9572	1.0000	1.0000
2	0.9463	0.9913	0.9943
3	0.9344	0.9843	0.9847
4	0.9284	0.9778	0.9945
5	0.9241	0.9646	0.9869
6	0.9223	0.9553	0.9571
7	0.9093	0.9539	0.9948
8	0.9058	0.9484	0.9519
9	0.9057	0.9425	0.9625
10	0.8764	0.9319	0.9559

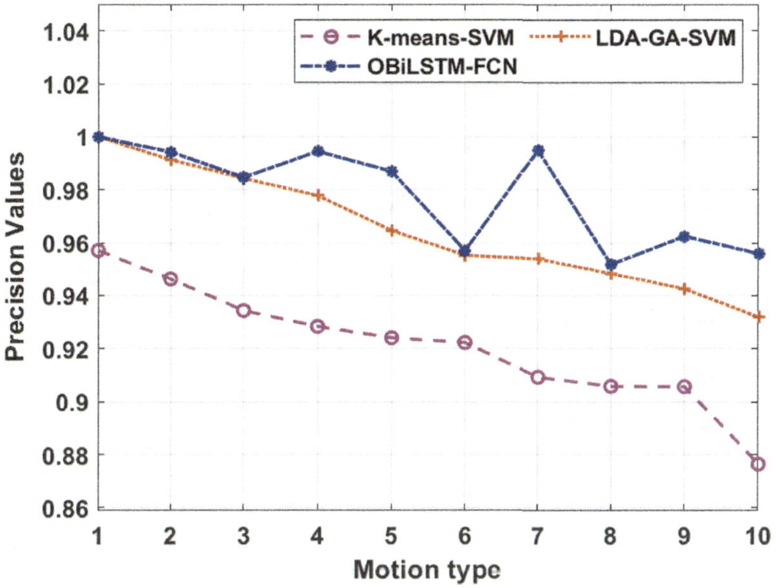

Fig. 3 Precision analysis of OBiLSTM-FCN model varying motion type

0.9646. Likewise, with motion type 10, the OBiLSTM-FCN technique has accomplished a higher precision of 0.9559 whereas K-means-SVM and LDA-GA-SVM techniques have demonstrated lower precision of 0.8764 and 0.9319.

Table 2 and Fig. 4 offer a comparative accuracy analysis of the OBiLSTM-FCN approach with other manners. The outcomes showcased that the OBiLSTM-FCN algorithm has reached effectual outcomes with superior accuracy values compared

Table 2 Accuracy analysis of OBiLSTM-FCN model with varying motion type

Motion type	Accuracy		
	K-means-SVM	LDA-GA-SVM	OBiLSTM-FCN
1	0.9116	1.0000	1.0000
2	0.9023	0.9992	0.9999
3	0.9379	0.9831	0.9963
4	0.9583	0.9802	0.9992
5	0.9076	0.9729	0.9789
6	0.9208	0.9615	0.9715
7	0.9565	0.9577	0.9850
8	0.9139	0.9507	0.9656
9	0.8910	0.9503	0.9530
10	0.9151	0.9497	0.9866

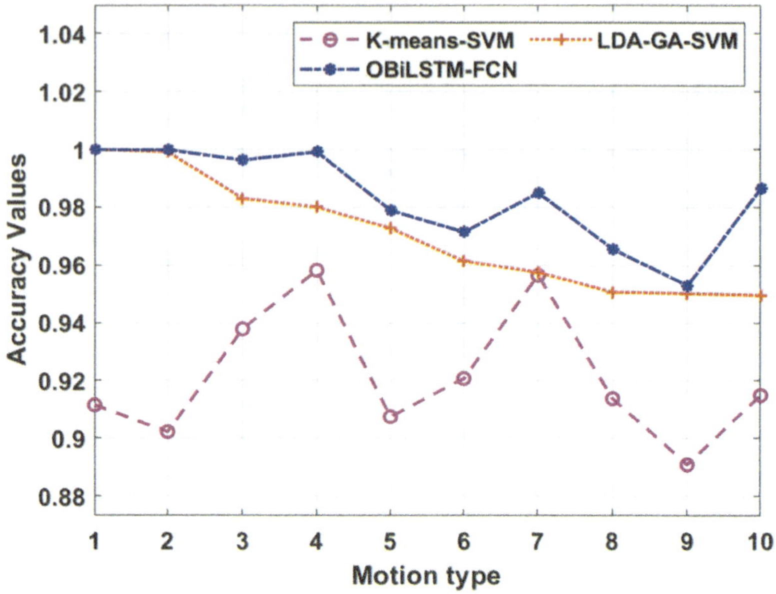

Fig. 4 Accuracy analysis of OBiLSTM-FCN model varying motion type

to other manners. For instance, with motion type 2, the OBiLSTM-FCN system has reached an improved accuracy of 0.9999 whereas K-means-SVM and LDA-GA-SVM techniques have achieved a reduced accuracy of 0.9023 and 0.9992. Followed by, with motion type 5, the OBiLSTM-FCN method has obtained an increased accuracy of 0.9789 whereas K-means-SVM and LDA-GA-SVM methodologies have resulted to lower accuracy of 0.9076 and 0.9729. Besides, with motion type 10, the OBiLSTM-FCN algorithm has accomplished a superior accuracy of 0.9866 whereas K-means-SVM and LDA-GA-SVM methods have outperformed reduced accuracy of 0.9151 and 0.9497.

Table 3 and Fig. 5 give a comparative specificity analysis of the OBiLSTM-FCN method with other systems. The outcomes exhibited that the OBiLSTM-FCN approach has system has gained effective outcomes with higher specificity values compared to other manners. For instance, with motion type 2, the OBiLSTM-FCN method has obtained an increased specificity of 0.9683 whereas K-means-SVM and LDA-GA-SVM techniques have gained a reduced specificity of 0.9249 and 0.9485. At the same time, with motion type 5, the OBiLSTM-FCN technique has attained a higher specificity of 0.9311 whereas K-means-SVM and LDA-GA-SVM systems have resulted to lower specificity of 0.9160 and 0.9307. Lastly, with motion type 10, the OBiLSTM-FCN algorithm has accomplished a maximal specificity of 0.9320 whereas K-means-SVM and LDA-GA-SVM methods have exhibited minimal specificity of 0.8321 and 0.9053.

Table 3 Specificity analysis of OBiLSTM-FCN model with varying motion type

Motion type	Specificity		
	K-means-SVM	LDA-GA-SVM	OBiLSTM-FCN
1	0.9377	0.9490	0.9774
2	0.9249	0.9485	0.9683
3	0.9245	0.9463	0.9532
4	0.9241	0.9314	0.9342
5	0.9160	0.9307	0.9311
6	0.9073	0.9259	0.9513
7	0.8828	0.9223	0.9902
8	0.8995	0.9108	0.9568
9	0.8944	0.9078	0.9600
10	0.8321	0.9053	0.9320

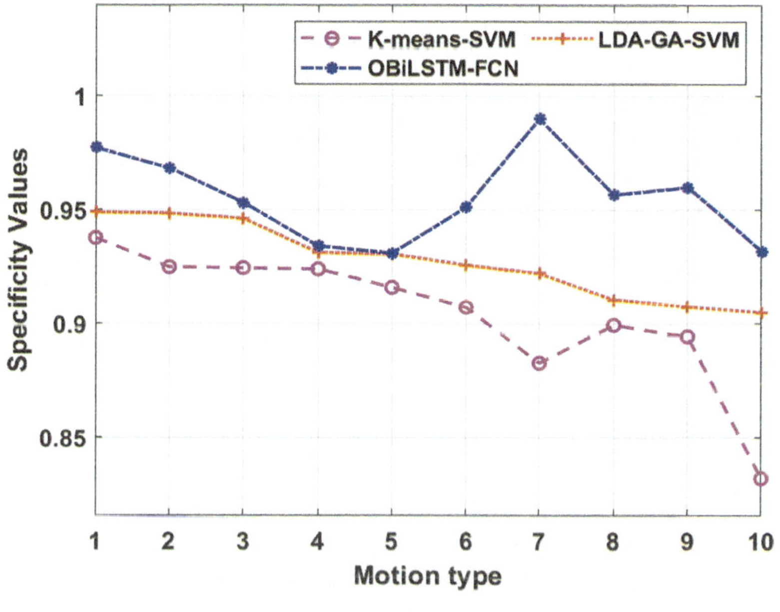

Fig. 5 Specificity analysis of OBiLSTM-FCN model varying motion type

Table 4 and Fig. 6 provide a comparative sensitivity analysis of the OBiLSTM-FCN manner with other approaches. The outcomes outperformed that the OBiLSTM-FCN method has attained effective results with the maximal sensitivity values compared to other algorithms. For instance, with motion type 2, the OBiLSTM-FCN methodology has reached a superior sensitivity of 0.9896 whereas K-means-SVM and LDA-GA-SVM algorithms have gained a minimum sensitivity of 0.9635

Table 4 Sensitivity analysis of OBiLSTM-FCN model with varying motion type

Motion type	Sensitivity		
	K-means-SVM	LDA-GA-SVM	OBiLSTM-FCN
1	0.9674	0.9666	0.9695
2	0.9635	0.9645	0.9896
3	0.9521	0.9558	0.9982
4	0.9320	0.9545	0.9733
5	0.9257	0.9524	0.9916
6	0.9185	0.9522	0.9853
7	0.8147	0.9305	0.9783
8	0.8350	0.9173	0.9942
9	0.8578	0.9162	0.9715
10	0.8561	0.8906	0.9022

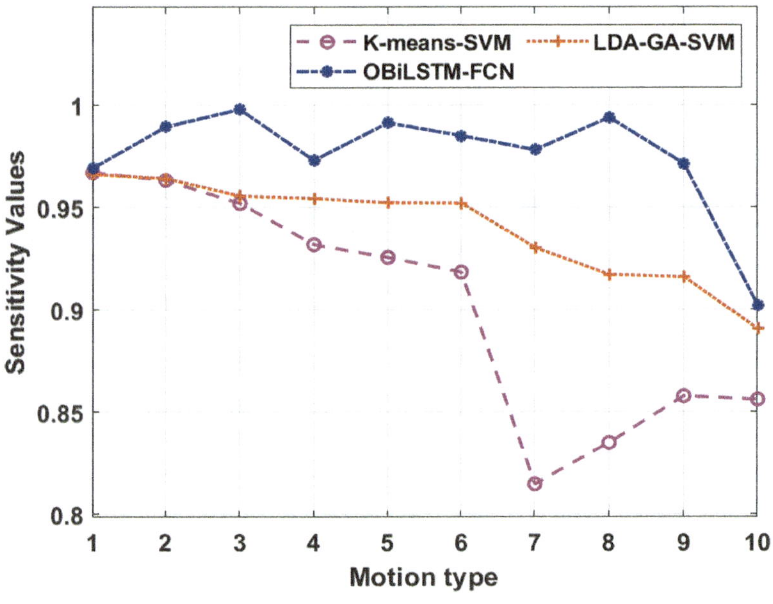

Fig. 6 Sensitivity analysis of OBiLSTM-FCN model varying motion type

and 0.9645. Along with that, with motion type 5, the OBiLSTM-FCN system has reached an enhanced sensitivity of 0.9916 whereas K-means-SVM and LDA-GA-SVM schemes have resulted in decreased sensitivity of 0.9257 and 0.9524. In addition, with motion type 10, the OBiLSTM-FCN manner has accomplished a higher sensitivity of 0.9022 whereas K-means-SVM and LDA-GA-SVM methodologies have demonstrated lower sensitivity of 0.8561 and 0.8906.

4 Conclusion

This study has designed an effective OBiLSTM-FCN model for human motion recognition in VR environment. The proposed OBiLSTM-FCN model comprises different processes namely feature extraction, classification, and hyperparameter optimization. Firstly, kernel based LDA approach is employed as a feature extraction technique. Besides, the BiLSTM-FCN technique is applied as a recognition model to determine human motions. Lastly, the Adam optimizer is applied to optimally tune the hyperparameters involved in the BiLSTM-FCN model. The performance validation of the OBiLSTM-FCN model take place and the resultant values portrayed the improved performance over the other compared methods. In future, metaheuristic algorithms can be incorporated for the hyperparameter optimization of the DL models.

References

1. Lun, R., Zhao, W.: A survey of applications and human motion recognition with Microsoft Kinect. Int. J. Pattern Recognit Artif Intell. **29**(05), 1555008 (2015)
2. Sulaiman, S., Tahir, N.M., Shah, A.M.M., Hussain, A., Samad, S.A.: Human motion analysis using virtual reality. In: 2007 5th Student Conference on Research and Development, pp. 1–4. IEEE (2007, December)
3. Qiu, S., Fan, X., Wu, D., He, Q., Zhou, D.: Virtual human modeling for interactive assembly and disassembly operation in virtual reality environment. Int. J. Adv. Manuf. Technol. **69**(9–12), 2355–2372 (2013)
4. Hu, F., Hao, Q., Sun, Q., Cao, X., Ma, R., Zhang, T., Patil, Y., Lu, J.: Cyberphysical system with virtual reality for intelligent motion recognition and training. IEEE Trans. Syst. Man Cybernet. Syst. **47**(2), 347–363 (2016)
5. Aleotti, J., Caselli, S.: Grasp recognition in virtual reality for robot pregrasp planning by demonstration. In: Proceedings 2006 IEEE International Conference on Robotics and Automation, 2006. ICRA 2006, pp. 2801–2806. IEEE (2006, May)
6. Kyan, M., Sun, G., Li, H., Zhong, L., Muneesawang, P., Dong, N., Elder, B., Guan, L.: An approach to ballet dance training through ms kinect and visualization in a cave virtual reality environment. ACM Trans. Intell. Syst. Technol. (TIST) **6**(2), 1–37 (2015)
7. Makris, S.: Virtual reality for programming cooperating robots based on human motion mimicking. In: Cooperating Robots for Flexible Manufacturing, pp. 339–353. Springer, Cham (2021)
8. Seth, A., Vance, J.M., Oliver, J.H.: Virtual reality for assembly methods prototyping: a review. Virtual Real. **15**(1), 5–20 (2011)
9. Zhang, N., Qi, T., Zhao, Y.: Real-time learning and recognition of assembly activities based on virtual reality demonstration. Sensors **21**(18), 6201 (2021)
10. Jiang, F., Yang, X., Feng, L.: Real-time full-body motion reconstruction and recognition for off-the-shelf VR devices. In: Proceedings of the 15th ACM SIGGRAPH Conference on Virtual-Reality Continuum and Its Applications in Industry, vol. 1, pp. 309–318 (2016, December)
11. Fu, Q., Zhang, X., Xu, J., Zhang, H.: Capture of 3D Human Motion Pose in Virtual Reality Based on Video Recognition. Complexity, 2020 (2020)
12. Fazeli, H.R., Venkatesh, S.K., Peng, Q.: A virtual environment for hand motion analysis. Procedia CIRP **78**, 127–132 (2018)

13. Kritikos, J., Zoitaki, C., Tzannetos, G., Mehmeti, A., Douloudi, M., Nikolaou, G., Alevizopoulos, G., Koutsouris, D.: Comparison between full body motion recognition camera interaction and hand controllers interaction used in virtual reality exposure therapy for acrophobia. Sensors **20**(5), 1244 (2020)
14. Sivasamy, M., Sastry, V.N., Gopalan, N.P.: VRCAuth: continuous authentication of users in virtual reality environment using head-movement. In: 2020 5th International Conference on Communication and Electronics Systems (ICCES), pp. 518–523. IEEE (2020, June)
15. Liu, L., Jiao, Y., Meng, F.: Key algorithm for human motion recognition in virtual reality video sequences based on hidden markov model. IEEE Access **8**, 159705–159717 (2020)
16. Zhou, M.: Feature extraction of human motion video based on virtual reality technology. IEEE Access **8**, 155563–155575 (2020)
17. Zhang, F., Wu, T.Y., Pan, J.S., Ding, G., Li, Z.: Human motion recognition based on SVM in VR art media interaction environment. HCIS **9**(1), 1–15 (2019)
18. Khan, M., Wang, H., Riaz, A., Elfatyany, A., Karim, S.: Bidirectional lstm-rnn-based hybrid deep learning frameworks for univariate time series classification. J. Supercomput. **77**(7), 7021–7045 (2021)
19. Zhao, Y., et al.: Applying deep bidirectional LSTM and mixture density network for basketball trajectory prediction. Optik **158**, 266–272 (2018)